Housing, Health and Well-Being

T0316358

Housing is a social determinant of health and this book aims to provide a concise source of the theory and evidence on safe and healthy housing to inform students, academics, public and environmental health practitioners, and policy-makers, nationally and internationally.

The book reviews the functions of housing and its relationship with the health and well-being of residents. It examines the implications of failures to satisfy those functions, including the potential impact on individuals, households, and society. It assesses options directed at avoiding, removing, or reducing threats and at promoting healthy indoor environments, particularly for the most susceptible and vulnerable members of society. It is essential reading for students, academics, and professionals within the areas of environmental health, public health, housing, built environment, social policy, housing policy, health policy, and law.

Stephen Battersby, MBE, is an environmental health practitioner, independent consultant, and advisor.

Véronique Ezratty, MD, is a medical doctor and an environmental health risk assessor at the Service des Etudes Médicales (SEM) of Electricité de France (EDF), Paris, France.

David Ormandy is a visiting academic specialising in housing and health, and is attached to Warwick Medical School, University of Warwick, UK.

Routledge Focus on Environmental Health
Series Editor: Stephen Battersby, MBE PhD, FCIEH, FRSPH

Housing, Health and Well-Being

Stephen Battersby, Véronique Ezratty and David Ormandy

Routledge
Taylor & Francis Group

LONDON AND NEW YORK

Chartered Institute of
Environmental Health

First published 2020
by Routledge
4 Park Square, Milton Park, Abingdon, Oxon OX14 4RN
605 Third Avenue, New York, NY 10017

First issued in paperback 2023

Routledge is an imprint of the Taylor & Francis Group, an informa business

© 2020 Stephen Battersby, Véronique Ezratty and David Ormandy

The right of Stephen Battersby, Véronique Ezratty, and David Ormandy to
be identified as authors of this work has been asserted by them in accordance
with sections 77 and 78 of the Copyright, Designs and Patents Act 1988.

British Library Cataloguing-in-Publication Data
A catalogue record for this book is available from the British Library

Library of Congress Cataloging-in-Publication Data
A catalog record has been requested for this book

ISBN: 978-1-03-257012-9 (pbk)
ISBN: 978-1-138-09698-1 (hbk)
ISBN: 978-1-315-10510-9 (ebk)

DOI: 10.1201/9781315105109

Typeset in Tmes New Roman
by Deanta Global publishing services, Chennai, India

Publisher's Note
The publisher has gone to great lengths to ensure the quality of this reprint but
points out that some imperfections in the original copies may be apparent.

Contents

Series preface

This series, Routledge Focus on Environmental Health, is now maturing and is no longer just an initiative. The aim remains the same: to explore environmental health topics, traditional or new, and sometimes contentious issues, in more detail than might be found in the usual environmental health texts.

We want to encourage readers and practitioners, particularly those who might not have had work published previously, to submit proposals, as we hope to be responsive to the needs of environmental and public health practitioners. I am very keen that this is seen as an opportunity for first-time authors and, as such, would urge students (whether at first- or second-degree level) to consider this an avenue for publishing findings from their research. Why, for example, should the hard work that has gone into a dissertation lie in an unread manuscript on a library shelf?

Our hope remains that this is a dynamic series, providing a forum for new ideas and debate on current environmental health topics. So if you have any ideas for monographs in the series please do not be afraid to submit them to me as series editor via the email address below.

I have always encouraged new authors and for environmental health practitioners on the front line to "get published", writing from their experiences of trying to protect public health. Setting down in writing some analysis of what worked in practice, what was successful, and what wasn't and why, can provide useful insights for others working in the field. It is not just an exercise in gathering CPD hours, but it can provide a useful method of reflection and aid career development, something that anyone who considers themselves a professional should do. Furthermore, all too often the work of EHPs goes unrecorded and unremarked and, with the demise of the *Journal of Environmental Health*, I am pleased to be working with Routledge to provide this opportunity to offer another route for practitioners to change this.

It is intended that this series should not take a wholly "technical" approach but that it should also provide an opportunity to consider areas of practice in a different way, for example looking at the social and political aspects of environmental health in addition to a more discursive approach on specialist areas.

We recognise that "environmental health" can be taken to mean different things in different countries around the world. I know that *Clay's Handbook of Environmental Health* has chapters that might not be relevant to some practitioners in different countries, nevertheless EHPs are a key part of the public health workforce wherever they practise. So, this series will enable a wider range of practitioners and others with a professional interest to access information and also to write about issues relevant to them. The format means a relatively short production time so contents will be more immediately available than in a standard textbook or reference work.

Forthcoming monographs are likely to cover such areas as licensing, environmental health in South Africa, and air pollution and health. That does not mean we have no need of further suggestions, quite the contrary, so I hope readers with ideas for a monograph will get in touch via Ed.Needle@tandf.co.uk or Patrick.Hetherington@tandf.co.uk.

Stephen Battersby MBE PhD, FCEIH, FRSPH
Series Editor

Preface

For any country in the world, the dwelling where you live affects both your physical and mental health. It is the very reason why, in the UK 100 years ago, the "Addison" Housing Act was needed and heralded the first massive house-building programme of council housing, helped by government subsidies, resulting in the construction of 500,000 houses within 3 years. So, even though the term was not used then, it was recognised that housing is a "social determinant" of health. This work aims to provide a concise source of the theory and evidence on safe and healthy housing and to inform students, academics, public and environmental health practitioners, and policy-makers, nationally and internationally.

A dwelling (be it a house or flat or some other structure) and its condition are crucial to the health and well-being of the occupiers. But the dwelling should be more than just the physical structure that provides shelter from the elements, it should permit the establishment of a secure home environment. The home is where we bring up families, socialise with friends, and have our own space, where it is possible to relax, keep our possessions safe, and take refuge from the rest of the world, allowing our households to flourish. It is where we also spend most of our time. The right home environment protects and improves health and well-being and helps to prevent physical and mental ill health.

As Public Health England has pointed out, poor, unsuitable, and precarious housing has a negative effect on our physical and mental health, particularly for older people, children, disabled people, and individuals with long-term illnesses. Good quality housing is critical to health.

This monograph reviews the functions of housing and its relationship with the health and well-being of residents. It examines the implications of failures to satisfy those functions, including the potential impact on individuals, households, and society. It assesses options directed at avoiding, removing, or reducing threats, and at promoting healthy indoor environments.

David Ormandy
Véronique Ezratty
Stephen Battersby

Part 1
Introduction

In 1968 a World Health Organization publication stated that "as a society advances, the rise in its cultural level and standard of living is accompanied by an immense increase in the demands made upon the home in which most a person's life is spent".[1] This continues to be true, but the speed and assortment of changes are increasing. The predicted effects of climate change will create new stresses on dwellings and their occupiers. These include what were seen as extreme events, such as heat waves, cold events, very high winds, and torrential rainfall leading to flooding, and so on; the changes in energy supplies and its use, such as the proposed banning of gas space heating in the UK; and the promotion of smart technology to control our energy use. Alongside this are the sociological changes such as housing becoming the workplace as well as the "home".

Housing is recognised as a social determinant of the health and well-being of individuals and households. Living in an inadequate dwelling that fails to meet the needs of occupiers has potentially serious consequences for health, well-being, and life chances.[2,3] Dwellings should provide a safe and healthy environment; providing shelter where the household can relax after work and school, be a place to enjoy family life, and feel safe; and should allow (and encourage) connection with the community, promoting social inclusion. Not only should dwellings not have a prejudicial effect on physical and mental health, they should have positive effect – good housing can lead to better health.

Housing and poverty are strongly connected and the likelihood of living in an inappropriate, unsatisfactory dwelling is greater among those on low incomes, particularly those below the poverty line. This in turn diminishes the opportunities for children and the young. It is not just a matter of the condition of the dwelling, but a number of interrelated issues. Households with low incomes can struggle to afford the rent or mortgage repayments, the local taxes, the charges for services (water, sewerage, etc.), and energy costs. Such problems are particularly exacerbated in energy inefficient dwellings, where the cost of trying to attain and maintain safe indoor temperatures can lead unacceptable decisions such as "heat or eat". The result is often poor diet, exposure to low indoor temperatures, increased risk of dampness and mould, self-imposed social exclusion, stress, days off work, and days off school.

There is also the potential increase in overheating in both older and new dwellings. With the anticipated changes to climate, heat waves are likely to become more common. In many temperate countries, dwellings were not, and are not currently, designed to protect against the consequential possibility of overheating. Adapting existing dwellings to provide protection can be difficult, and building codes for new (yet to be built) dwellings are not always taken into account. However, exposure to high indoor temperatures has serious health implications for the elderly, the very young, and other vulnerable members of the population.

The objective of this monograph is to provide a review of the theory and evidence relating to safe and healthy housing and what healthy housing should look like. It is intended to be a resource for students, academics, public and environmental health practitioners, housing professionals, and policy-makers. While the majority of sources used are based on housing in the developed countries, the principles are international. As environmental health and housing professionals have to be able to assess housing at an individual level, and also contribute to and inform policies and strategies, this monograph is wide-ranging in its approach.

The range of functions of housing are reviewed, and the relationship between housing conditions and the health and well-being of residents discussed. This includes examining the implications of a failure to satisfy the basic functions, with a focus on the potential impact on individuals, on households, and on society. Options for avoiding, removing, or reducing threats are also discussed, with particular emphasis on those who may be more susceptible.

As well as a place of residence, housing (specifically, a dwelling) is a financial and social asset. It is a financial asset (an investment) to the owner, either as a place to live, or as a source of income (from rent). Dwellings are also social assets, both locally and nationally.

This monograph however will concentrate on the physical structures and environment – i.e., the dwelling and the neighbourhood – as these provide the backdrop and the framework to housing. Deficiencies in these interfere with the development of the social structures – the home and the community. It also highlights the connections and relationship between the various aspects and disciplines involved in housing – those concerned with the structure, those concerned with the sociology, and those involved with health and well-being.

Notes

1 Goromosov MS (1968) The Physiological Basis of Health Standards for Dwellings, *Public Health Papers* 33, World Health Organization, Geneva.
2 Marmot Review Team (2011) *The Health Impacts of Cold Homes and Fuel Poverty*. Friends of the Earth and the Marmot Review Team, London. https://friendsoftheearth.uk/sites/default/files/downloads/cold_homes_health.pdf.
3 https://www.who.int/sustainable-development/publications/housing-health-guidelines/en/.

Part 2

Aspects of a dwelling

The relationship between a dwelling and the household involve four main aspects – adequacy; compatibility; affordability; and security of occupation. These are discussed below.

Adequacy

A dwelling should be properly designed, constructed, and maintained so as to provide an optimum safe and healthy environment for the occupiers. As such, it will reflect cultural, economic, climatic, geographic, and other local and regional factors. It should also take into account known and predicted extreme events that may affect the locality, such as floods, earthquakes, heat waves, and cold events.

Each element of, and facility in, a dwelling should satisfy its particular function(s) and do so without interfering with the occupation or with the functions of other elements or facilities. Such elements, modules, and facilities include:

- elements such as foundations, walls, roof, windows, doors, etc., that make up the structure and fabric of the building;
- modules such as bedrooms, living rooms, kitchens, bathrooms, etc.; and
- facilities and amenities including water supply, energy (for space and water heating, lighting, and ventilation), passive ventilation, sanitation, personal and domestic hygiene, etc.

(The functions are discussed in Part 3.)

Adequacy should also recognise the basic needs of a household, such as the need for "family life", including space for members of the household to get together, and the need for privacy such as spaces for personal necessities (personal hygiene, etc.), and somewhere to be alone and complete personal tasks (study and homework, etc.).

Compatibility

A dwelling may be adequate in itself, but unsuitable for particular house-holds or individuals. Examples are where a dwelling is too small for a household of a particular size, resulting in crowding and interference with family and individual life; or an apartment in a "walk-up" block that is unsuitable for a family with small children, or a household that includes an individual with restricted mobility. In such cases, while there is nothing wrong with the dwelling, it is the incompatibility that is the problem, the mismatch between the dwelling and the occupiers.

Affordability

There will be various costs associated with a dwelling and these should be within the means of the household. Unaffordable housing has been shown to be detrimental to mental health, particularly for low- to moderate-income households.[1]

Where the occupying household is purchasing the dwelling, there will be the repayment of instalments for a loan (mortgage) and associated interest. Such repayment is vital to avoid foreclosure (repossession).

The dwelling may be rented, or, where the physical structure is owned, the land on which it stands, or to which it is moored, may be rented (land-less owner-occupation). Payment of rent is vital to retain occupation. Where a dwelling is an apartment within a block there will be service charges towards maintenance and general services (cleaning, etc.).

Assessing affordability must also include other "living costs", including food, clothes, furnishings, and miscellaneous costs associated with human life and contentment. There are other regular or irregular financial outgo-ings associated with any housing – "running costs". For example, these can include local taxes and standing charges for services (energy, water, sewer-age, etc.). Where a dwelling is rented, theoretically, an element of that rent should go towards the maintenance of the structure (although this may be through service charges). Where a dwelling is subject to a loan or mortgage, the cost of maintenance etc. will fall to the household.

Affordability is an important consideration that feeds through to other aspects. If "affordable" is defined as a percentage of "market rents", then when inflation results in the cost of housing outstripping wage and income increases, so-called "affordable" rents become, in reality, "unaffordable". Rent control linked to local wage levels may be a one method of ensuring affordability, but this may have other consequences.

Low-income occupiers, and those with a disability (depending on the definitions), may qualify to receive financial support towards their housing cost (usually rent) or the cost of adaptations. In effect, taxpayers are subsidising the rents paid, which means that a personal benefit is a form of subsidising housing, rather than the other option of managing the cost of housing.

Assistance towards the cost of services is less certain, and, in some cases non-payment for energy or water may result in disconnection (although, in some countries this is prohibited either completely or during certain, colder, seasons).

Security of occupation

For a household to settle and establish a home, there needs to be some certainty that the right to occupy that dwelling will continue for a reasonable period. While there will be some conditions attached, such a right to occupy should be sufficient to give peace of mind and assurance.

What amounts to a reasonable period will vary, but it should reflect the needs of the household (students, for example, may not need or want a long-term tenancy beyond the academic year, while a household with school-age children will need a right to occupy for a longer period to allow for continued attendance at the same school).[2]

- A study in 2018 looking at tenancies in the private sector in England with those in other jurisdictions, highlighted several differences and reported on examples of good practice, i.e., where security was recognised as important for the well-being of occupiers.[3] In England the length of tenure for most tenants in private accommodation is 6–12 months, and is considerably shorter than that in other jurisdictions. For example, tenancies in Italy are typically for 4–6 years and in Germany tenancies are for no fixed term. English tenancies are renewed at the volition of the landlord, while in other jurisdictions, such as New Zealand and Australia, tenants can automatically renew the tenancy once the previous agreement has expired, provided that the landlord has failed to give notice that the tenancy will be ending.
- Landlords in England can obtain possession by giving just two months' notice without giving any reason, once the first six or twelve months of the agreement has ended. Other jurisdictions require longer notice periods and place various restrictions and limitations on a landlord's ability to obtain possession. For example, many of the states in Australia

require a landlord to give a valid reason to obtain possession before the end of the fixed term. In Germany the landlord's resumption of possession is likely to be subject to lengthy and costly court proceedings and, upon resumption, the landlord must also return the tenant's security deposit plus additional interest, which provides financial support to tenants in their relocation.

The neighbourhood

A dwelling cannot be seen in isolation from the immediate housing environment. For an apartment within a block it will include the common or shared elements of the structure giving support (the foundations), and protection from the local climate (the roof and the external walls). It will also include the party walls and ceiling/floor sandwiches that provide separation between the individual dwellings.

For all dwellings, the means of access is important. For a dwelling within a block this will include the common means of access from public areas, and any common stairs (or lifts/elevators). For individual (separate) dwellings it will include the access from public or shared areas.

Households need access to services and facilities, such as work, schools, places of worship, places of entertainment, and day-to-day supplies (food and household goods). There should also be places for relaxation (such as green spaces) and for association with other local residents, which will have a positive effect on mental health. The neighbourhood should be designed and maintained to encourage exercise, and provide general feelings of safety.

As Public Health England has recognised, green space and good quality surroundings affect well-being and mental health by reducing stress and sadness. There are also physical benefits from green infrastructure, for example improved air quality, less noise pollution, and reduced risks from flooding or heat waves. There are also benefits to active users of these spaces, whether that's physical recreation or through children interacting with nature.[4]

The community

For the household, the community is the social, cultural, and economic structure created by all those within a neighbourhood. As well as the residents, it includes all those providing services through the workplaces, the schools, the places of worship and entertainment, and the shops.

Notes

1 See Bentley R et al. (2011) Association Between Housing Affordability and Mental Health: A Longitudinal Analysis of a Nationally Representative Household Survey in Australia. *AJE* 174(7): 753–760.
2 For example, Becker Cutts D et al. (2011) US Housing Insecurity and the Health of Very Young Children. *AJPH* 101(8): 1508–1514.
3 Woods S (2018) *Comparing the Rights of Private Sector Tenants in England with Those in Other Jurisdictions: Examples of Good Practice.* Centre for Human Rights in Practice, University of Warwick, Coventry.
4 Landscape Institute (2016) *Public Health and Landscape – Creating Healthy Places Position Statement.* https://landscapewpstorage01.blob.core.windows.net/www-landscapeinstitute-org/migrated-legacy/PublicHealthandLandscape_CreatingHealthyPlaces_FINAL.pdf.

Part 3

Functions of the dwelling

The assessment of any dwelling should be based on the use of the structure and the potential effect (particularly, any negative effect) on that use and on the users – that is the residents. This also depends upon a sound basic understanding of the construction.[1] The principle behind the assessment is to determine whether the structure and facilities enable the premises to be used as a dwelling. That is, does the dwelling satisfy the basic physiological and psychological requirements for household life and comfort, and provide protection from the environment, from infection, and from accidental injury.

It is important to recognise that households may move from one dwelling to another, so a dwelling should be suitable for occupation by households across the spectrum of possible lifestyles and needs. While a relatively young employed individual may make few demands on a dwelling, the demands and use made by a household with young children, or a household comprised of an elderly couple, will be very different and much greater.

Households with children will have particular requirements from a dwelling. The first environment for a child is within the dwelling. Children need to be protected from cold and/or heat, they need somewhere safe and clean as they learn to crawl, walk, play, and explore. Households that include an elderly individual also need somewhere safe, but for different reasons. Like children, they need to be protected from cold and heat, but they also need ready access to facilities when mobility becomes limited.

To determine whether the structure and facilities pose a threat, it is necessary to understand the particular function(s) of each element and each facility. It is when an element of a facility is either missing or fails to fulfil its function(s) that there may be a threat; i.e., it is the effect of defects that is important. A defect or deficiency should not be judged by the extent or cost of repair or replacement, but the threat it produces. For example, a loose window catch may cost little to replace, but until replaced a small child could easily open the window and be at risk of falling out.

The functions of some of the main elements and facilities of a dwelling and the potential threats from failure are discussed below. Before this discussion, a couple of points should be noted.

First, some functions of elements are obvious, but most elements serve several functions, some of which are less obvious but nonetheless important. For example, an internal door obviously allows access into a room, but it will also, when closed, complete the separation provided by the internal walls, allowing for different activities within different areas/rooms, allowing for different temperatures in different parts of the dwelling, and providing privacy. Some internal doors will also complete the compartmentation and help to limit the possible spread of fire and smoke.

Second, while the discussion below considers the general functions of elements, facilities, and rooms, the functions should take account of the locality (and the local geology and climate) and of cultural demands and needs.

The following discussion of some of the main elements and facilities of a dwelling is not intended to be exhaustive but is given as an indication of the approach.

Exterior

Foundations – These are intended to provide support and stability for the whole structure and the use to which the structure will be put. They should take account of the local geological conditions, of the local climate, and of any known or potential environmental hazards (such as earthquakes, storms, or flooding). At its worst, the failure of the foundations can result in the complete collapse of the whole building; less severe failure can cause settlement and fracturing of walls (particularly at door and window openings), damage to the integrity of damp-proofing, and displacement of floors.

As well as instability of the structure, fractures can result in water penetration and dampness internally.

External walls – These are the vertical erections that contain the dwelling. They provide support for the floors and roof as well as protection against the local climate, reducing to a minimum heat loss and heat gain, and preventing rain or precipitation penetration. They should also protect against the penetration of sound (noise).

There should be no cracks or other defects that could allow moisture to penetrate into the structure.

Some materials used for walls will draw water from the ground by capillary action. To prevent this, an impervious barrier (a damp-proof course) is necessary to prevent dampness affecting the structure and living space.

Some materials used for external wall construction do not provide adequate protection against driving rain and require a waterproof external finish (such as cement render or tiles). This can often provide a decorative finish as well as protection. Hairline cracking or defects to the protective outer finish will allow water to penetrate or cause water to be drawn into the wall structure by capillary action. Once behind render the water cannot escape outwards and can only drain down or dry out through the interior surface finishes.

The weakest points in any external wall are at door and window openings. The joint between the wall and the door or window frame must be tight and weatherproof to avoid draughts and water penetration.

The sill to window openings must be sloped so that any water from the glazing runs off, away from the window frame, and it must be provided with a drip or throating (a groove on the underside) to break the adhesion of the water so that it drips off the sill rather than running into the wall structure.

Internally, the wall surfaces should be smooth and capable of being readily kept clean (particularly wall surfaces in the kitchen, bathroom, and sanitary rooms or areas).

Roofs – The roof or roofs cover the upper surface of the dwelling and complete the enclosure of the dwelling. They must be constructed so as to support not only their own weight but the additional loads imposed by winds and accumulations of snow. As well as giving protection against the weather, they should provide thermal and sound insulation.

In many regions traditional roofs are finished with tiles or slates. Cracked, loose, or missing tiles or slates will allow water to penetrate into the building. Distorted or sagging pitched roofs suggest that the support roof timbers are defective, perhaps as a result of being of inadequate strength, or of fungal or insect attack.

Generally, flat roofs require more maintenance than pitched roofs, and, because they are less easily checked, any deterioration may go unobserved. Felt- or asphalt-covered roofs should be provided with some protection from solar heat (usually in the form of reflective chippings) to limit expansion and contraction, which may increase the rate of deterioration. Uneven flat roof coverings will retain water, again increasing the rate of deterioration.

Where the roof covering is penetrated by service pipes or chimneys the joint between the roof covering and the pipe or chimney must be watertight.

Rainwater goods – Surface water running off the roof is collected and safely taken away from the building by the rainwater goods. These include the guttering immediately below the eaves and any fall or downpipes.

Defective rainwater goods may discharge the contents onto a wall and cause damage to the wall and penetration into the dwelling.

Although not rainwater goods, in areas where there is likely to be snow, there should be guarding at the eaves of sloping roofs to protect against accumulations slipping off.

Means of access – There should be safe and unhampered access to the dwelling from public footpaths. For dwellings within a block, the means of access include the common passageways, stairs, and lifts/elevators. There should also be safe and unhampered access to any amenity space for clothes drying, for play and recreation, and to any outbuildings associated with the dwelling.

Paths and yards should be laid so as to be self-draining and to avoid ponding of surface water. Uneven, cracked, and holed yards and paths will interfere with the safe passage of the occupiers and will collect surface water.

External doors – These provide for access into and out of the dwelling, but also complete the weather protection, privacy, and security provided by the structure. To ensure adequate weather protection, the door should fit closely to the frame, and should stop and be held securely closed by the door lock.

Ill-fitting doors with badly adjusted locks will allow draughts and water penetration, as well as providing inadequate security against unauthorised entry.

Interior

Internal walls – These are the vertical structures that divide the dwelling into separate rooms and areas, enabling different activities to be carried out in each room; they also give privacy to individual members of the household and enable personal tasks to be carried out in private, without interference to other occupants. They may or may not provide some structural support.

The walls should provide some thermal and sound insulation, enabling different parts to be heated or cooled to different temperatures, and limiting noise nuisances within the dwelling. The walls should also limit the spread of fire within the dwelling.

Internal surfaces of all walls should be smooth and even so that they can be decorated and easily kept clean. This is particularly important in the kitchen, bathroom, and WC compartment where hygiene should be easy to maintain. (Wall plaster and ceramic wall tiles are not mere decorative finishes, but provide the smooth, even, and, in the case of tiles, impervious, finish necessary.)

Plaster affected by damp will not remain properly decorated and cannot be easily kept clean and hygienic. Cracked and loose plaster, and cracked and broken tiles, are difficult to clean, and can harbour dirt and insect pests.

Skirting boards seal the joint between the wall surface and the floor (bridging and protecting the gap between the plaster and the floor surface).

Internal doors – These complete the separation provided by the internal walls, while allowing access between different parts of the dwelling; and when closed they complete the separation provided by the internal walls. They should also give privacy and prevent draughts when closed.

Doors, particularly to the kitchen (even if not certified fire-rated), should help limit the spread of fire through the dwelling, and, to be effective, they should be close-fitting to the frame and stop. Doors should be provided with the necessary door locks and handles to ensure that they are held close-fitting and are capable of being readily opened and closed. Ill-fitting doors with gaps greater than 8 mm will allow draughts and allow smoke and flames to pass.

Ceilings – Ceilings provide the vertical separation within the dwelling, separating one floor from another and the top floor from any roof void.

They should provide sound and thermal insulation, limit the spread of fire within the room and between floors, and should have a smooth and even surface, capable of being decorated and easily kept clean.

Floors – Floors should be capable of supporting the weight of both the occupiers and the furniture likely to be used in a dwelling. They should be even (avoiding any trip hazards), and easy to keep clean, and should complete the separation between the room and the room or space below. As for the ceilings, they should provide sound and thermal insulation.

Windows – Provide natural lighting to the interior of the dwelling and give occupants a view of the outside environment. Openable lights to windows provide a means of natural ventilation. They also complete the structure and exterior, providing weather protection and some sound and thermal insulation (particularly double- and triple-glazed windows).

Opening lights should be capable of being opened and securely closed. When closed they should be close-fitting against the frame to provide security, and to prevent draughts and water penetration. Cracked or holed glazing will allow draughts and water to penetrate, cannot be properly cleaned, and will negate any thermal or noise insulation.

Facilities

Kitchens – The kitchen or kitchen area is where food is prepared, and some domestic activities carried out. The kitchen or kitchen area should be capable of being readily cleansed and kept clean, with smooth surfaces; and those surfaces adjacent to fittings and amenities should be impervious.

Sinks are used for the preparation of food, for the cleansing of cooking and eating equipment and utensils, for the washing of clothes, and for personal washing. The material of the sink, and the internal surface in particular, should be smooth, impervious, and capable of being easily kept clean and hygienic. The sink should be close-fitting to adjacent wall surfaces, with the joints watertight to ensure splashed water runs safely back into it. It should be properly connected to a trapped waste pipe capable of safely carrying the waste water out of the dwelling and into the drainage system, and it should be provided with a supply of water (either separate hot and cold, or a mixture tap (faucet)).

Drainers should be smooth, impervious, securely fitted, and self-draining into the sink. Drainers should be close-fitting to adjacent wall surfaces and to the sink, with the joints watertight to ensure water runs back into the sink.

Food preparation surfaces should be smooth, impervious, and capable of being easily kept clean and hygienic. Joints between the preparation surface and adjacent walls should be watertight.

Food storage cupboards should be capable of being thoroughly cleaned and should be provided with close-fitting doors capable of being readily opened and closed. They should be designed and maintained so as to prevent access for pests. There should be space for refrigerators and freezers; although the space need not be within the kitchen, it should be readily accessible.

There should be facilities for cooking food. These facilities should be of adequate size for the intended household, and they should be properly and safely designed. In the case of cooking facilities fuelled by gas, proper consideration should be given to the provision of adequate ventilation to provide sufficient air for combustion as well as the removal of the waste products of combustion; and the burners should be properly maintained to ensure completed combustion.

There should be space or facilities for the drying of clothes during cold and wet weather, whether within or adjacent to the kitchen.

This could be a properly designed cabinet with provision for heating, or space for an electric tumble dryer. The provision for drying clothes should include the means of ventilation to remove moisture-laden air quickly and safely out of the dwelling (for example, an extractor fan, ventilation of the clothes drying cabinet to the external air, or a vent for connection to the ducting from a tumble dryer).

There should be a lockable (childproof) cupboard for the storage of potentially poisonous substances, such as cleaning products.

Bathrooms and WC (sanitary) compartments – Baths, showers, and wash hand basins are used for personal washing. The internal surfaces should be smooth, impervious, and capable of being easily kept clean and hygienic. They should be close-fitting to adjacent wall surfaces, with the joints watertight to ensure splashed water runs safely back into the facility. Baths and showers should be securely fitted to ensure that they are capable of supporting their own and the occupant's weight, and a wash basin should be securely fitted to ensure that it can support the weight of an occupant leaning on it. The facilities should be properly connected to trapped waste pipes that can safely carry the discharged water out of the dwelling into the drainage system, and they should be provided with supplies of water.

WC (water closet) basins should have smooth, impervious internal and external surfaces and be self-cleansing. They should be connected to a properly working, flushing cistern that is provided with a supply of water, and is also properly connected to a drainpipe capable of safely carrying the waste out of the dwelling and into the drainage system. The basin should be provided with a water seal to prevent foul air escaping from the system. It should be securely fixed and capable of carrying its own weight and that of the user. It should be fitted with a hinged seat. All the cisterns, basins, pipes and drains should be watertight. There are alternatives to WCs, such as chemical toilets and composting toilets. Chemical toilets retain the waste in a tank or cassette, which should be charged with a chemical to prevent decomposition and smells until the tank or cassette is emptied. Composting toilets usually use a dry material to cover the waste while allowing it to decompose, until the tank/container is emptied.

There should be a lockable (childproof) cupboard for the storage of potentially dangerous substances, such as medicines.

Space heating – In temperate and colder regions there will be a need for appliances to raise the indoor temperature to safe levels (i.e., within the thermal comfort zone).[2] This provision should be of an appropriate type having regard to the design and construction of the dwelling, the materials used in the construction, and the location of the dwelling.

A solid fuel heating system requires adequate storage provision for the fuel, and the means of disposal for the waste ash. Such systems require more attention from the occupiers than systems using other fuels, particularly as they are less easily started and stopped. A system fuelled by oil also requires adequate storage provision.

Open fires are a relatively inefficient means of space heating compared to other forms. An open fire will be at best only around 40–50 per cent efficient (with 50–60 per cent of the heat being wasted up the flue), a modern, high-efficiency solid fuel boiler will be up to 75–80 per cent efficient. Gas fires and gas-fired boilers will be 75–85 per cent efficient, and electric heaters will be up to 100 per cent efficient.

In addition to the type of provision for space heating the distribution of the heat is important. Radiators sited under window openings will reduce draughts and provide greater comfort. A forced-warm-air system will reduce the risk of stagnant air pockets in room corners, while high-temperature radiant heaters will leave parts of rooms cold and relatively unheated even though occupants are warm and comfortable.

(NB: Electric socket outlets and gas points are not provision for space heating, but sources of energy or fuel that may be used for different types of space-heating facilities.)

Cooling – In hot and temperate regions some provision for cooling the interior of the dwelling is needed. In temperate regions this may be achieved through passive means, while in hotter regions it will involve appliances such as air conditioning.

Passive means include blinds, awnings, and *brise-soleil*, which protect against solar heat gain. (In the UK, windows usually open outwards, which makes fitting shutters problematic. In most European countries, windows open inwards, allowing shutters to be closed from inside and windows to be left open for ventilation.)

Air conditioning is intended to reduce the temperature of warm air and may also improve the purity of the indoor air. Such equipment uses energy and, because of any filters, requires regular maintenance and cleaning.

Air quality – Activities within dwellings will generate impurities, and ventilation is necessary to reduce these.

Natural ventilation (via openable windows and fixed ventilators) may be sufficient, depending on the quality of the external ambient air and prevailing wind direction. Any natural ventilation should avoid both draughts and excessive heat loss during winter periods. In kitchens and bathrooms ventilation should be provided by mechanical means to ensure moisture-laden air is quickly and safely taken out of the dwelling close to points where high levels of water vapour are generated, such as cookers, showers, baths, and sinks).

In many regions, there is a need for drying clothes indoors. This is an obvious source of moisture, and a cabinet vented to the outside provides the best option, but the general ventilation should take clothes drying into account.

Air conditioning will usually provide a means of mechanical ventilation, and it may incorporate the means for heat recovery too (i.e., the means to avoid heat loss by recovering heat from the air before discharging it).

Notes

1 See, for example, Marshall D, Worthing D et al. (2013) *The Construction of Houses*, 5th edn. Routledge, Oxon.
2 Ormandy D, Ezratty V (2012) Health and Thermal Comfort: From WHO Guidance to Housing Strategies. *Energy Policy* 49: 116–121.

Part 4

Research and evidence

When an environmental health or other health professional is advocating action to address housing conditions, whether at the policy level or on an individual basis, it is necessary to support arguments with evidence. This evidence can in part be obtained from reported research, however researching to provide evidence to support the apparently obvious association between housing conditions and health can be difficult. The main problem is that unsatisfactory housing is invariably linked with other social, physical, and/or economic, confounding factors. Unfortunately, it is not sufficient to assert the connection, and it is sometimes seen as necessary to try to untangle these factors, at least in part, so as to justify and inform policies and actions.

The various research methodologies used to investigate the relationship have their own strengths and weaknesses, some practical and others evidential. Perhaps unfortunately, as "health" is a focus, the models tend to be based on those developed and used in the medical sphere, which can be strictly controlled. For those working on housing, health has a wider meaning and it can be unhelpful to rely on the medical research model, where the controls required in such research models are almost impossible to replicate in the housing field.

The following research methods are discussed below, giving brief outlines and considering their pros and cons:

- descriptive or cross-sectional studies
- case control studies
- longitudinal studies
- intervention studies
- experimental studies
- modelling
- systematic reviews.

Descriptive or cross-sectional studies

These simply try to document events, are usually geographical (i.e., area based) and consist of a description of the housing and of the health of the residents. They are cross-sectional as they deal with a specific limited time. They may be quantitative, focusing on a particular feature of a dwelling, or qualitative, canvassing self-reported health and satisfaction.

The usefulness of the findings will be influenced by the size of the study (the number of dwellings or individuals), and by the housing features and health conditions taken into account.

Several aspects of these types of studies should be considered. First, there is an interrelationship between the many physical features of a dwelling that may not be included in the study. An example is where the relationship between the form of heating and respiratory conditions is investigated without considering other relevant factors such as the thermal insulation of the dwelling, the provision for ventilation, and the location of the dwelling (e.g., whether it is close to a busy road). There are also social or behavioural factors that can influence the results, such as the available income and the composition of the household.

It should also be recognised that correlation does not imply a causal link. For example, there may be a strong correlation between ice cream sales and the incidence of sunburn; but it is another factor that is the link. Any suggested causal link between housing conditions and health also should be plausible biologically.

There are examples of useful and important cross-sectional studies. One is the World Health Organization's LARES project,[1] useful because of the comprehensive data collected on households and on dwellings. It is also important because of the amount of data collected – 8,539 individuals, in 3,382 dwellings, from 8 European cities. Other examples deal with the impact on mental health from the loss of home following a flood.[2] In these studies the causal link between the event and the health of those affected is direct and cannot be considered merely a correlation.

The benefit of this approach is also that those living in the dwellings are engaged in the research, and their own description of their health status can be as valuable as, if not more so than, any independent assessment. It can also take account of their mental health, an aspect that is often forgotten. This approach can also be useful for practitioners as an adjunct to their routine work.

Case control studies

These usually begin with a health condition, such as asthma. Two groups of participants would be recruited, one being those diagnosed as suffering from asthma, and the other a group of similar individuals without asthma. The study would then compare the housing conditions of each group.

To try to avoid confounding factors, the information on housing should be relatively comprehensive and, ideally, the housing should be similar.

Such studies, while more complicated to set up than cross-sectional studies, are likely to produce more convincing results.

Longitudinal studies

These studies are designed to follow a group of people, some of whom will have been exposed to environmental hazards (for these purposes, hazards related to housing) over a period. They are complex and will not provide results for several years.

However, there have been some examples that have given useful and important information. These include the Whitehall studies, which involved collecting data from a cohort of 17,500 male civil servants between the ages of 20 and 64; there have been several analyses using data that were originally collected in 1967, but the studies have continued since then.[3] While the Whitehall studies focused more on lifestyle and work, the HELIX Project has focused on combining the impact of environmental hazards that mothers and children have been exposed to and linking this with the children's development and growth.[4]

Both the exposure (to potential hazards) and the health outcome need to be well documented. In housing terms, the hazards will probably have preceded the study, but adequate records can provide the necessary information.

There is likely to be a loss (reduction in the number) of participants over the period of the study. The size of the loss depending on the timescale of the study. This means there should be sufficient numbers involved to avoid the results being compromised. Where the loss in participants is 20 per cent or more, the findings may be an underestimation.

Intervention studies

As the name suggests, these are studies where there a change has been made to housing (such as the upgrading of the energy efficiency), and the effect on the health and well-being of the residents assessed. The ideal housing intervention study will involve two groups, one being those whose dwellings are subject to the intervention, and the other those whose dwellings remain untouched. This is, however, difficult in the housing field and may raise ethical issues, as one group will be expected to remain in unimproved dwellings (for period of time). Nonetheless, there are possibilities, such as where there is a programme of improvement.[5]

These studies should not be confused with a "before-and-after" study. There are several problems with such studies. There will be the time difference between "before" the intervention and after the intervention, as well as possible negative impacts from the disruption caused by the intervention. Additionally the time difference may also mean that occupiers have changed and this will reduce the valid sample size.

These studies can provide reliable results but are expensive and difficult to set up and coordinate.

Experimental studies

Experimental studies are where the impact is hypothesised from the results of non-domestic studies. Examples of such studies are those dealing with the exposure of animals to some volatile organic compounds. There have been some studies (with ethical approval) using human volunteers, such as those involving exposure to formaldehyde and allergens.[6] However, exposure to radon is based on that experienced by a group of miners.

One problem relates to dose/exposure and extrapolating this to the domestic environment. However, such studies do provide the basis for a precautionary approach.

Modelling

This is the process of generating a model as a representation of some or a range of phenomena. Modelling is often used to predict the possible effects of climate change, or the effect of energy usage in buildings.[7] It can also be used to predict the effect of interventions, such as energy upgrading of dwellings.

Systematic reviews

These are reviews of existing literature that relate to specific and well-defined topics. They include searches to identify, appraise, and synthesise research evidence. They are usually considered to be best source of research evidence.

These are usually more thorough than a straightforward literature review and may cover published and unpublished material, including grey literature (see below, p. 25).

Non-research based evidence

There are other sources of evidence that can support the design, construction, and adaptation of dwellings. First, there are the obvious, self-evident potential hazards. The use of child-safety catches or restrictors on windows will, undeniably, reduce the possibility of a child opening a window and falling out – the idea of the need for a research study to confirm this is absurd. The same can be said for automatic cut-off devices for gas appliances (shutting off the gas supply if the flame is extinguished).

Some elements of dwelling designs are based on laboratory testing, including the design of staircases (such as the treads, risers, and handrails), the height of kitchen worktops, and the positions of door handles and light switches. Again, there seems little need for research studies to confirm the safety and benefits of the results of such testing. While these developments in practice suit the majority of the population, there may be a need to consider whether the general results suit the needs of those with functional limitations or disabilities, and whether special provision should be made for children.

Sources of information

Research studies are reported in numerous academic publications aimed at a wide range of trades and disciplines. These include journals focusing on medical/health matters, architecture, engineering, acoustics, building construction, housing, and public health. Fortunately, most are now listed in online searchable libraries. Most of the research study reports included in these academic journals will have been subject to peer review, a process that attempts to ensure strict adherence to research protocols and clarity of reporting.[8]

As well as research reports, there are publications that cover developments made in the relevant industry or by independent research bodies – the so-called "grey literature". These will not have been subjected to peer review but should still provide useful information on the developments. A relative recent publication that brought together evidence from both research reports and grey literature is *Review of Health and Safety Risk Drivers*.[9] This provides summaries of risks and preventative measures on an extensive range of potential hazards in dwellings and other buildings such as offices.

Evidence

There are continuing investigations into the relationship between aspects of housing and health (both physical and mental), so the summaries presented here will have a short |shelf life", and it is necessary to check for recent research reports. The discussions that follow are based on a mixture of practical codes and international standards, research evidence, and the obvious or self-evident potential hazards.[10]

Although this section separates the evidence and guidance under different headings, it is important to recognise the interaction between each of the topics. There is, for example a relationship between the thermal environment, dampness, and indoor air quality – changes to one will have an effect on the others. It is not unreasonable to assume, although there has been little research into this, that there could also be a "cocktail effect" in that some hazards in housing have a greater effect when linked with other hazards or deficiencies (such as low indoor temperatures and falls).

Internal thermal environment

Ideally, for temperate regions the internal thermal environment should be between 18 °C and 24 °C; this is generally referred to as the thermal comfort zone, although this range is about health as well as "comfort".[11] This thermal comfort temperature range has been adopted by most European countries.

The health impact of exposure to temperatures below 18 °C (excess cold) is usually delayed, often by several weeks and can have fatal consequences as a result of respiratory and cardiovascular conditions. As temperatures decrease below 18 °C the potential impact on health increases in severity. The body's reaction to low temperatures includes thickening of the blood and hypertension, leading to an increased risk of cardiovascular or cerebrovascular events, such as heart attacks and strokes. Respiratory stress starts at around 16 °C and cardiovascular stress when the temperature falls below 12 °C. The lower the temperature falls, hypothermia (a drop in the body's core temperature) becomes a possibility.

In 2014, Public Health England carried out a review of the literature relating to exposure to low indoor temperatures and reported that temperatures of 18 °C pose minimal risk to sedentary individuals wearing suitable clothing.[12]

Low indoor temperatures are also associated with other threats to health such as dampness and mould associated with respiratory conditions. In addition, the cold seasons brings with it other potential threats to health, such as respiratory disease epidemics that may contribute to an increase in morbidity and mortality, although the relationship with indoor air temperatures is not clear. Excess winter deaths (EWDs) are seen as the headline for exposure to low temperatures, but it is not clear whether these deaths are primarily related to low indoor temperatures or other factors, such as thermal shock caused by going from the heated indoors to the cold outside. However, there seems to be a stronger relationship between indoor temperature and blood pressure than with outdoor temperature.

It has also been suggested that there is an association between extreme low temperatures and birth defects, and some evidence that living in a cold home has negative impacts on mental health in all age groups.

Heat-related health impacts, unlike those for cold, appear to occur within a few days of exposure, and some heat-related deaths may be explained by mortality displacement, although this is less clear with cold-related deaths.

The health impacts from heat (i.e., in temperate regions, above 24 °C) can vary by region and by community. It also seems that individuals and populations can become more resilient to heat over time. The perception of what constitutes uncomfortably high temperatures (and so personal discomfort) varies depending on local and regional climatic conditions and perhaps sociocultural factors. For example, a Finnish study found that an ambient temperature of 26 °C was considered "hot", and that cold discomfort started at around 22 °C. Such temperatures may be considered normal or even cool in other regions.

The health effects of exposure to high temperatures can start with dehydration, then overheating, and heat exhaustion. It also increases the risk of heat-stroke, and problems such as respiratory and cardiovascular hospitalisations and deaths.

There is strong evidence of an association between high temperatures and an increased risk of stillbirths and shortened gestation, and a suggestion that, to reduce the risk of sudden infant death syndrome (SIDS), infants, with appropriate clothing and bedding, should sleep in rooms at temperatures of between 16 °C and 20 °C.[13]

(See also below, p. 48, on extreme events, where temperatures far exceed the normal range.)

Dampness and associated mould growth

There are several sources for dampness. These include:

- Moisture is present from the evaporation (drying out) of water used in the construction or rehabilitation process. A large quantity of water is used in the construction process of a traditional building. Even with "dry construction systems" there will be some water used in construction. Such construction moisture dries out slowly and, as it does so, the relative humidity within the building will be high, meaning condensation will be more likely. That drying out process will be slowed if the building is occupied.
- Capillary action will draw water from the ground to affect the lower parts of walls and ground floors. This is generally referred to as rising damp. Ideally there should be a damp-proof course to the walls and a membrane to the floors to prevent the capillary action.

- Penetrating dampness is when moisture enters the building through defects and weaknesses to the external skin of a dwelling – through external walls, particularly around the joints between window and door openings and the walls, or through defects to the roof.
- Traumatic dampness results from leaks to water pipes, water storage tanks, or waste pipes or drainpipes.
- There can be condensation and high levels of relative humidity. Condensation, as mentioned in relation to drying out following construction or rehabilitation, occurs when warm moisture-laden air is cooled by a cold surface. Moisture is naturally generated by the domestic and biological activities of the household. Unless this moisture is safely dealt with (usually by ventilation) relative humidity levels and condensation can become a problem.

Moisture passing through parts of the structure will dissolve salts from the material. These salts tend to inhibit mould growth (but not always), although decoration and the structural material will be damaged. The moisture may contribute to indoor humidity levels (although only to a minor extent) and will reduce the thermal insulation values of the structure. Thus, high levels of humidity and condensation, neither of which will have absorbed salts, are linked to mould growth and an increase in the dust mite populations, and house dust mite detritus and mould spores (including timber-attacking fungi) are potent allergens.[14]

Exposure to concentrations of dust mite detritus over a period will result in the sensitisation of atopic individuals (i.e., those with a genetic tendency to sensitisation). In addition, exposure to high concentrations over long periods can also sensitise non-atopic individuals. Once sensitised, exposure to relatively low concentrations of the allergen can trigger allergic symptoms, including rhinitis, conjunctivitis, eczema, cough and wheeze, and repeated exposure can lead to asthma and asthmatic events.

The spores of many moulds and fungi (including timber-attacking fungi) can be allergenic. The spores of some species can be carcinogenic, toxic, and cause infections. Those on immune-suppressant drugs can be vulnerable to fungal infection, and very high concentrations of spores can colonise the airways of susceptible individuals.[15]

(See also below, p. 48, on extreme events dealing with flooding.)

Indoor air quality

There are both biological and non-biological pollutants. Biological pollutants include mould spores, house dust mites, and some viruses (particularly in crowded dwellings). Mould spores and dust mites are dealt with in the above section, Dampness and associated mould growth.

There is a wide range of potential health-threatening non-biological sub-stances that can pollute indoor air, including:

a)   asbestos
b)   biocides
c)   carbon monoxide and fuel combustion products
d)   uncombusted fuel gas
e)   volatile organic compounds (VOCs)
f)   lead dust
g)   radiation (in the form of radon gas).

There will also be pollutants from the external air, which may include particulates (such as $PM_{2.5}$ and PM_{10}) and oxides of nitrogen (NOx).[16]

Asbestos – Asbestos fibres were incorporated into a wide range of building products used in some countries before 1980, particularly where some fire resistance was required, but also in some other finishes such as floor tiles. Although some of these products were in positions unlikely to be disturbed, in some non-traditional dwellings, products containing asbestos were in positions vulnerable to damage and disturbance. Indoor air concentrations of asbestos fibres in most dwellings are such as to present minimal risk to health. It is only when materials containing asbestos are damaged that air-borne asbestos fibre levels could be higher and potentially dangerous. Work such as plumbing and rewiring may lead to the release of asbestos fibres, though fortunately exposure from such activities is likely to be infrequent and for only a short period.

Biocides – These are products used to try to protect timber and as a treat-ment for infestations, fungi, and moulds. Fumes emanating from them are a risk health if inhaled, however, the production of gases lasts for only a relatively short period (approximately two to three weeks depending on ventilation) and so the main risk is the period following use (both post-construction or rehabilitation).

Carbon monoxide, etc. – Carbon monoxide (CO), oxides of nitrogen (NOx), sulphur dioxide (SO_2), and smoke (containing particulates includ-ing $PM_{2.5}$ and PM_{10}), are products associated with the combustion, or incomplete combustion, of carbon-based fuels including gas, oil, and solid fuels used for heating and cooking. CO is a colourless, odourless, and extremely toxic gas. As blood haemoglobin has a greater affinity for CO than for oxygen, inhalation of this gas reduces the ability of the blood to take up oxygen. At high concentrations CO can cause unconsciousness and death. At lower concentrations, it causes a range of symptoms includ-ing headaches, dizziness, weakness, nausea, confusion, disorientation, and fatigue, symptoms that are easily mistaken for influenza and depression.

Carbon monoxide may also impair foetal development. The elimination of CO from the blood is relatively short (often less than eight hours), which means that diagnosis of CO poisoning can be difficult, unless a blood test is taken within hours of exposure. Those most vulnerable to CO poisoning include foetuses, infants, the elderly, and people with anaemia or heart or lung disease. NOx, and particularly nitrogen dioxide (NO_2) affects the respiratory system, damaging the lining of the airways. At low levels it may cause narrowing of the airways in asthmatics and may exacerbate reactions to allergens such as house dust mites (see above, p. 28). Asthmatics are therefore more vulnerable than others, particularly if also exposed to other airborne allergens. Exposure to NOx may also increase susceptibility to lung infections.

Sources of the various products of combustion can leak or be discharged into the dwelling as a result of defective or blocked flues. Open fires, even where there is an apparently adequate flue, can be a source of SO_2 and smoke when winds cause a down-draught. SO_2 and smoke have been shown to cause respiratory conditions, particularly bronchitis and breathlessness. In some regions, solid fuel cooking stoves or open cooking facilities can discharge SO_2 and smoke directly into the kitchen area.

Uncombusted fuel gas – The various gases used for fuel can, where they leak into a dwelling, displace the normal air to such an extent that the occupants are unable to obtain sufficient oxygen to breathe. This can result in asphyxiation. In the UK, mains gas is primarily methane, which is less dense than air. Bottled gas used in rural areas is liquefied petroleum gas (LPG), either propane or butane, which is denser than air. Both mains gas and LPG are odorised to give distinctive smells and so alert users to the danger of escaped gas. (The difference in densities is important for the positioning of gas sensors.) Very young children (those aged under 5 years), the elderly, and pregnant women are the most vulnerable.

Volatile Organic Compounds (VOCs) – These cover a wide group of organic chemicals (including formaldehyde) that are gaseous at room temperature, and are found in a wide variety of materials in the home, including adhesives (used for floorings) and paints. VOCs are likely to be released for a period following construction or rehabilitation, and adequate ventilation is necessary to purge the dwelling. While there is little evidence that the individual VOCs that may be found in dwellings have long-term health effects, some may cause short-term irritation and allergic reactions to the eyes, nose, skin, and respiratory tract. Allergy sufferers, such as asthmatics, are most vulnerable, and may react to VOC exposure at levels below those that would affect others. Formaldehyde can be a particular problem, although individual sensitivity varies.[17]

Lead – This is a heavy metal, which, if ingested, accumulates in the body. It has toxic effects on the nervous system, cognitive development, and blood production. Continual exposure at low levels has been shown to cause mental retardation and behavioural problems in children. Lead is readily absorbed from the intestinal tract, especially in children, and its absorption is enhanced by dietary deficiency of iron and calcium. The main source of lead is lead-based paint that has deteriorated. The most prevalent risk is intelligence quotient (IQ) deficiency in children, rather than acute poisoning. Even with relatively low levels of lead in blood, there are indications that it affects the IQ of children. The highest risk group is that of young children aged 0–3 years because of lead's potential effect on neurological development, and because, physiologically, they take up lead more readily. Children may also ingest lead from paint (pica) or dust. Pregnant women and foetuses have also been identified as a risk group, mainly in relation to levels of lead in water. The elderly are more susceptible to health effects than younger adults, because, as part of the ageing process, lead may be released from bone changes, and toxic effects may be observed from relatively low lead exposures. (Lead has been and is the driver behind healthy housing in the USA.[18])

Radiation – The main sources of radiation within dwellings are radon gas and its daughters, primarily airborne, and radon that may be dissolved in water from private water supplies (wells and boreholes). Radon gas and its decay products (which are also radioactive) pose a threat of lung cancer. Radon decays rapidly and the resulting products can very quickly attach themselves to particles in the air, which if inhaled, can be deposited in the lungs where the process of radioactive decay continues. The particles emitted during decay can cause cells lining the lungs to be genetically mutated, and initiate cancer, or facilitate a process already initiated by other carcinogens. The risk related to radon increases with dose and duration of exposure, and the risk relates to the local geology, some areas having no risk of exposure (maps give the likelihood of exposure[19]). Those most at risk are persons over 64 years who have had lifetime of exposure and are smokers. While there have been suggestions that there are adverse health effects from electromagnetic fields (EMFs), there is no clear evidence of a risk to health from low-level exposure to the EMFs normally found in the domestic environment.

Space and privacy

Frequently, space within dwellings focuses on the density (the number of individuals), in particular the space available for sleeping, although sometimes including a reference to living space. Lack of space and crowded conditions have been linked to a number of health outcomes, including psychological distress and mental disorders, especially those associated with a lack of privacy and childhood development.[20] There is evidence that crowding can result in an increased in heart rate, increased perspiration, reduction of tolerance, and a reduction of the ability to concentrate. Crowded conditions are also linked with increased hygiene risks, an increased risk of accidents, and spread of infectious disease.[21] There appears to be no particular age group that is particularly susceptible compared with others, although it seems logical that those who spend the most time at home will be affected most, typically the elderly, the very young, the mobility impaired, and their carers.

There should be space to provide for social interaction between members of the household, as well as space(s) allowing for private time away from other household members. Conversely, too much space may lead to a sense of physical and social isolation, particularly for single and elderly persons.

Personal space and privacy needs are important for the individual members of the same household as well as for individuals or households sharing rooms and/or facilities. These needs vary reflecting both individual and cultural perceptions. Adolescents may need more space than the elderly; small children need at least as much space as an adult. The need for privacy begins to develop from the age of 8 years and will be fully formed during puberty.

For households of more than two individuals, there should be a physical separation of living, cooking, dining, and sleeping areas. Bedrooms should be large enough to be usable for sleeping and for study or relaxing away from the other members of the household. There should be sufficient bedrooms having regard to the numbers likely to be accommodated in the dwelling. (Different countries have proposed "standards" dealing with space and crowding, and these are described and compared in Part 5.)

There should also be a living area of sufficient size for the household. Indoor and outdoor play and recreation spaces are necessary for households with children, and any outdoor play space should be readily visible from within the dwelling and safely separated from public and neighbouring areas.

To provide for adequate privacy for the user, baths or showers should be sited in a bathroom and each sanitary closet should be sited in a bathroom or separate compartment, both provided with a lockable door. (See section below on "Personal hygiene and sanitation", p. 35.)

Security and fear of crime

Occupiers need to be able to keep their dwelling secure against unauthorised entry and to maintain the dwelling as defensible space. Residents also need to feel safe when walking around the immediate housing environment.

The potential health effects of feeling insecure within one's own home include:

a) the fear of a possible burglary occurrence or recurrence;
b) the stress and anguish caused by a burglary; and
c) injuries caused to occupants by an intruder (aggravated burglary).

The emotional impact is greater for burglaries where there is successful entry to the dwelling. Fear of burglary or crime outdoors usually arises from knowing someone who has been burgled or robbed, or from publicity about such crimes in the local area.[22] It seems that while elderly people may be more fearful of walking on the streets after dark, they are less anxious about burglary than other age groups.

Lighting and outlook

Inadequate light or lack of an outlook can cause:

a) stress and depression from a lack of natural light or the lack of a window with a view;
b) disturbance by intrusive artificial external lighting at night;
c) eyestrain from glare and a lack of adequate light (natural or artificial);
d) flicker caused by certain types of artificial light, which can result in discomfort and may cause photoconvulsive reactions in those who are susceptible.

Poor light (natural and artificial) can mean that the elderly (whose vision may be slow to adjust to changes in light levels) and those with impaired vision are more likely to be unable to detect potential hazards.[23]

The amount of natural light depends on the shape, position, and size of windows and the layout of rooms. Windows, adequate in themselves, can be obstructed (overshadowed) externally by other buildings or by trees. Basement dwellings may suffer from poor natural lighting, while attic dwellings may not provide a reasonable outlook, leading to feelings of isolation. External lighting (street lights and security lighting) should be sited to avoid annoyance and sleep disturbance.

Noise

The sensitivity and tolerance to noise varies among individuals, including those living together. Tolerance may in part be determined by age, gender, working status, lifestyle, and personality. And, while noise levels can be measured (in decibels), there may be differences in what sources and types of noise individuals find offensive.

Noise is likely to be tolerated from neighbours in the daytime, as well as some traffic noise and normal or expected deliveries (such as mail), but the same noises may cause annoyance during the night-time. People are less likely to tolerate noises that are unusually loud, continuous, or thought to be unnecessary or inconsiderate. Noises such as shouting and those associated with violence are poorly tolerated.

A dwelling should be constructed so as to protect the occupants from noise penetrating from outside the dwelling, this includes protecting occupants from ordinary domestic noise from one dwelling entering another (such as from an attached dwelling in a row or terrace, or in a block), and from traffic or other ambient external noise.[24]

As well as causing sleep disturbance, noise has been shown to have a physical effect, and can affect heart rate and blood pressure.[25]

Domestic hygiene

A dwelling should be designed, constructed, and maintained so as to be readily cleansed and kept clean. This is particularly important in areas where food is kept and prepared and in areas for personal hygiene (bathrooms and WC compartments) although it applies throughout a dwelling. There should be a supply of water, preferably hot or temperature-controlled, to facilitate cleaning.

Walls and ceilings should be smooth and even to enable them to be easily cleaned and decorated. Floors should be smooth and even so that they can be easily kept clean. All internal surfaces should be smooth, even, and free from cracks and crevices that may allow entry by, or give harbour to, pests. Joints between walls and floors and between walls and doors and windows should be effectively sealed.

Service ducting, roof spaces, and underfloor spaces should be effectively sealed off from the living area. However, there should be means of access to these spaces for treatment in case of any infestation.

Pests create a risk of cross-contamination and infection, carry disease, and can infect food and surfaces.[26,27] Structural defects, such as broken vents to suspended timber floors, can enable the entry of insects, rodents, or other pests to the dwelling. Insect pests, such as flies, can contaminate foodstuffs, others may attack humans, cause annoyance, and spread disease. Rodents are perpetual inhabitants of the sewers, from where they can readily gain access to drains and, unless prevented, they may gain access to dwellings where they may spoil large quantities of food both by gnawing and through indiscriminate fouling. They may also cause fires by damaging electric cables.

There should be suitable and sufficient provision separate from the living space within a dwelling for the storage of refuse awaiting collection or disposal. The storage provisions should be readily accessible to the occupants, should not cause problems of hygiene, nor attract or allow access to pests, and should be sited so as not to create a danger to children.

Personal hygiene and sanitation

Poor personal hygiene and sanitation can result in gastrointestinal conditions, ranging from mild stomach upsets through to severe dysentery and gastroenteritis. Dysentery and rotavirus infections, causing diarrhoea, are spread by the faecal-oral route. Even if the illness is contracted elsewhere, infection may be spread where personal hygiene cannot be maintained. Odours associated with poor personal hygiene can result in social isolation, and be a cause of stress.

There should be a sufficient number of baths or showers for the occupants. These should be connected to a supply of water at a controlled temperature or to supplies of hot and cold water, and connected to pipes to safely carry the waste water away from the dwelling, discharging it into the drainage system (sewers or storage/treatment tanks).

Baths or showers should be sited in a properly designed, heated, lighted, and ventilated bathroom, provided with a lockable door to give privacy. There should also be a sufficient number of wash hand basins with separate supplies of cold water and hot water, or water at a controlled temperature, and sited so as to facilitate use. Wash hand basins should be sited in the bathroom and, if there is a separate WC compartment, within that compartment.

Kitchen sinks may be used for hand-washing of clothes as well as for food preparation and for the washing-up of kitchen and eating equipment. Again, there should be separate supplies of cold and hot water, or water at a controlled temperature, over each sink. Each sink should also be properly connected to pipes that safely carry away the waste water from the dwelling.

Ideally, there should be space for a washing machine (with an appropriate power socket adjacent), and clothes drying facilities. The internal clothes drying provision should, ideally, be vented to the outside and provided with means of low-level heating.

Water closets should have a smooth and impervious surface and be self-cleansing. They should be connected to a flushing cistern provided with a supply of water, and also properly connected to a drain capable of safely carrying waste out of the dwelling and into the drainage system. The design of a water closet basin should incorporate a water seal to prevent odours and access by rodents. There are other forms of sanitary equipment such as composting closets, chemical closets, and water closets with macerators.

The sanitary facility should be located in a separate ventilated compartment or a bathroom, which should be of a hygienic design and construction. There should be a door to the compartment or bathroom capable of being locked (although, in an emergency, openable from the outside).

Bathrooms and sanitary compartments should be of sufficient space, giving functional space around each facility (space for ease of use).

Sinks, wash hand basins, baths, showers, bidets, and other water-using facilities must be properly connected to adequately sized waste pipes capable of safely carrying the waste water out of the dwelling and into the drainage system. Each waste pipe should incorporate a trap to provide a water seal of adequate depth to prevent draughts and foul air entering the dwelling.

Where waste water from personal, clothes, and food equipment (so-called grey water) is to be recycled, it should be stored in a tank outside the dwelling. Alternatively, it should be discharged into a properly constructed soakaway. Pipes carrying foul waste from sanitary closets must be air-tight to avoid leakage of the foul sewage or smells.

Water supply

Water is essential to sustain life. At normal temperature, with little or no exercise, an adult in temperate regions needs around 2.5 litres of fluid each day, but this will rise substantially in hot conditions. Mild dehydration is associated with fatigue, headaches, dry skin, constipation, bladder infections, and poor concentration.

Legionella can be dispersed into the air during the use of showers, and this, although rare, is the most likely route for transmission of legionnaires' disease in homes. *Legionella* thrive between 20 °C and 45 °C. To prevent *Legionella* growth, hot water needs to be maintained above 55 °C. This means hot water storage tanks should be set to store hot water at above 60 °C. (But, to avoid scalding, there should be thermostatic mixer valves at taps.) It should also be noted, that if hot water is used regularly and not stored for long periods, this reduces the risk of an infective dose of *Legionella*.

Filters in water supply systems, unless kept clean or replaced regularly, can allow pathogens to proliferate.

Water for drinking, cooking, washing, and laundry needs to be of high quality. However, water for flushing toilets can be of lower quality, such as reclaimed rainwater, recycled grey water, or bathroom waste water. An average two-person household will use around 110 litres of water per day. All dwellings should have at least one drinking water outlet.

Potentially dangerous features and events

The risk of physical injuries can be increased by various features in the dwelling, or the layout. Slips or trips resulting in a fall and a physical injury can occur in various locations. There are also potentially dangerous facilities and services that can cause physical injuries (or even fatalities). These include:

1) falls:
   a) slips and trips associated with floors, passages, and paths, including falls associated with trip steps (very small, unexpected, and indistinct changes in floor levels), thresholds, or ramps, where there is a change of level of less than 300 mm;
   b) slips and trips in bathrooms or shower rooms;
   c) slips, trips, and missteps associated with stairs, steps, and ramps where there is a change of levels of 300 mm or more;
   d) falls from one level to another, such as out of windows or from balconies and landings, where the difference is 300 mm or more, whether inside or outside;
2) cuts and strains, and collisions and entrapment (including injuries associated with collapse of part or all of the structure);
3) poisonings;
4) shock and burn injuries:
   a) shocks and burns associated with electricity;
   b) shocks and burns associated with hot surfaces, materials, exposed flames; and
   c) injuries or fatalities associated with uncontrolled fires;

while the outcome may be some form of physical injury of varying severity, the causes are different, and so are discussed separately.

Fall injuries

Falls can result in physical injury, such as bruising, fractures, head, brain, and spinal injuries.[28] The nature of injury is in part dependent on the distance of a fall, and in part dependent on the nature of the surface onto which the victim falls. Hard surfaces such as uncovered stone, concrete, or ceramic-tiled floors being more unforgiving than carpeted floors. As well as the surface of the landing site, the distance fallen will affect the outcome.

Cold impairs movement and sensation, and a lowered body temperature affects mental functioning, with the result that falls are more likely in the cold. The thermal efficiency of the dwelling (see above, p. 26, on thermal environment) is therefore relevant to fall hazards, as well as whether the incident occurs outdoors on a path, for example. It is likely that it is more hazardous using external paths in cold weather, irrespective of whether they are wet or icy.

Following a fall, the health of an elderly person can deteriorate generally, and the cause of death following an initial fall injury can be cardiorespiratory. This may include heart attack or pneumonia and may not necessarily result directly from the impact injury sustained at the time of the fall.

Floors, passages, and paths – The likelihood of a slip or trip occurring on what should be a relatively level surface is affected by how level and even that surface is, and its state of repair. Even small variations can increase the likelihood of a slip or trip. While falls on the level tend to result in relatively minor injuries compared with other falls, they occur more frequently.

The construction, evenness, inherent slip resistance, drainage (for outdoor path surfaces), and maintenance of the floor or path surface, all affect the likelihood of a slip or trip and the severity of the outcome. Other factors such as lighting, temperature, and distracting noise also have an effect.

The possibility of a slip occurring is affected both by the slip resistance of the floor surface and by the characteristics of any footwear. The type of floor covering will determine the final slip resistance. Slip resistance is worsened when a surface is damp or wet, which may be the result of a building deficiency, or be expected given the use of the area in question.

A lack of sufficient space to carry out tasks or manoeuvres may also increase the likelihood of a fall.

Bathrooms and shower rooms – The main cause of falls in bathrooms is slipping when getting into or out of a bath or shower unit. Whether a slip occurs will depend on various features, such as the slip resistance of the internal surfaces of baths and showers, and whether there are any handles or grab rails.

The position of taps, waste controls, and other bathroom controls can also affect both the likelihood of a slip or trip and the severity of the outcome. Inappropriate siting may force a user to reach awkwardly, increasing the risk of a slip, and the position may increase the severity of an outcome in the event of a fall. The position and direction of the opening of the door may also affect the likelihood of a fall.

Inadequate functional space (the space necessary for using the facility) immediately adjacent to the appliance may make it more difficult to use, increasing the likelihood of a fall. Inadequate lighting or glare can also increase the likelihood of a fall, as can a light switch remote from the doorway.

As noted above, cold impairs movement and sensation, and in bath and shower rooms the user will remove clothes before using the facility.

This means that a fall may be more likely in a bath or shower room that cannot be adequately heated, and the outcome (in part because of the lack of clothes) from a fall may be more severe.

To reduce the likelihood of a slip or fall, baths and showers should be stable and securely fitted, provide for slip resistance, and incorporate safety features such as handles or grab rails and the side positioning of taps and waste controls. The layout of a bathroom and of the appliances should allow for ease of use of each appliance, including sufficient functional space to enable users (and a carer assisting a child or another individual) to be able to undress, dry, and dress without increasing the likelihood of a fall.

Stairs, steps and ramps – These include falls associated with stairs or ramps within the dwelling and stairs and ramps outside the dwelling (whether internal in a block or external). It also includes falls over guarding (balustrading) associated with the stairs, steps, or ramps. Falls over guarding to balconies or landings are considered separately, as are falls associated with trip steps, thresholds, or ramps where the change in level is less than 300 mm.

The design of stairs and steps can increase the likelihood of a misstep. Variations in the dimensions of riser and/or going to a straight flight are likely to increase the possibility of missteps, although a change of rise or going linked with an obvious change in direction of a stair may mean that the user takes greater care, which can reduce the likelihood of a misstep. This suggests that intermediate landings (with or without a change of direction) can reduce the likelihood of a misstep.

As well as increasing the potential severity of outcome, a particularly long straight flight of stairs or long ramp may increase the chances of a misstep and fall. The likelihood of missteps is reduced where treads are 280–360 mm and the rise 100–180 mm respectively. Accidents are more likely where the pitch of stairs is more than 42 degrees, and, for ramps, the steepness of the slope is relevant to the potential for accidents. The shape and dimension of nosings affect the likelihood of an occurrence. In particular, nosings that project more than 18 mm may increase missteps.

Poor frictional quality of the surface of stair treads, and particularly of nosings, can increase slips and missteps. An accident is more likely to occur on stairs without carpet covering, including those stairs intended to be left uncovered. Although loose and ill-fitting carpeting will increase the chances of a misstep.

As well as providing a means of assistance when climbing or descending stairs, handrails are a safety device if there should be a misstep and can help prevent a fall. Handrails on both sides of the stairs provide the safest arrangement. Handrails should be sited between 900 mm and 1,000 mm measured from the top of the handrail to the pitch line or floor. They should be shaped so that they are easy to grasp and extend the full length of the flight. Where the is no wall to one or both sides of the stairs, guarding (e.g., balustrading) should be provided to prevent falls off the sides of stairs. The guarding should be designed and constructed so as to discourage children from climbing onto it. In addition, to prevent small children falling or becoming trapped, openings to stairs and guarding should not be large enough to allow a 100 mm diameter sphere to pass through.

Ideally, the headroom to stairs should be a minimum of 2,000 mm. However, where this is not possible (such as for loft conversions), the headroom could be reduced to 1,900 mm at the centre reducing to a minimum of 1,800 mm at the side.

Narrow stairs may cause problems in emergencies. Ideally, stair width should be at least 900 mm to allow stairs to be negotiated by a child and adult side-by-side. In residential blocks, stairs should be wider as they will be used as a means of escape.

Good lighting at the top and bottom of the stairs is important, allowing users to identify the first and last stair, and the dimensions of the stairs. Artificial lights and windows should be sited to avoid shadows or glare that may interfere with a user's vision. There should be switches or controls for artificial lighting at both the top and the foot of the stairs. As well as adequate lighting, to allow users to appraise the start and dimensions, there should be reasonable space at the top and bottom of stairs. Architectural features (including doors) that could create an obstruction can increase the chances of a fall. Projections and sharp edges on stairs and glass or radiators at the foot of stairs will increase the severity of an outcome from a fall.

The chances of a fall may be increased on spiral stairs where there is no inner handrail and where the width of the tread is less than 800 mm. Alternating tread stairs (paddle stairs) may also be hazardous, particularly in emergencies.

Uncovered external steps, which may become icy or wet, or are uneven and badly maintained, will increase the likelihood of a fall and the severity of the outcome.

Windows, balconies, and landings – For windows, the ease of opening, the distance they can be opened, the height of the sill, and the design of the opening light are all relevant to the possibility of a fall out of the window. The design of the windows, particularly those above ground floor level, should facilitate safe cleaning of the outer surface, and avoid the possible need to climb on furniture or a stepladder to clean them.

In some cases windows may be a means of escape in case of fire, such as windows to the first and perhaps second storey, or those giving access to a safe space. For these, the window should be readily openable in an emergency, but still be child-proof.

Windows that can easily be opened by a child will increase the possibility of a child accident, and contrariwise, windows that are difficult to open may increase the likelihood of an accident for an adult, as they may find it difficult to reach the catches (although this may be more relevant to strain injuries, discussed below, p. 42).

Safety catches reduce the possibility of a child being able to open a window unsupervised, and restrictors that limit a window opening to 100 mm will reduce the possibility of an accident involving a child. Internal sills should normally be at least 1,100 mm from the finished floor level to discourage children from climbing onto the sill. However, any restrictor should be easy to override by an adult in the event of fire, and, where a window is intended to serve as a means of escape in case of fire, then the bottom of the openable area may be 600 mm above the floor (a trade-off between the likelihood of a fall and fire safety).

Although 1,100 mm is an ideal height for an internal sill considering children, to allow views from a seated position and for wheelchair users the height of glazing above floor level should not be more than 800 mm. Where there is any glazing extending to within 800 mm of the floor level, it should be guarded or of safety glass. And, while there should be a catch to at least one window in a room accessible to wheelchair users, such a window should still be fitted with a restrictor.

Guarding (balustrading) should be provided to balconies and landings to prevent falls. Such guarding should be at least 1,100 mm high, designed and constructed so as to discourage children climbing, and also strong enough to support the weight of people leaning against it. There should be no openings to the guarding that would allow a 100 mm sphere to pass through.

The distance of a window opening or balcony above the adjacent ground will affect the severity of the outcome of a fall, as will the nature of the surface on which an individual will land. The greater the distance and the less forgiving the surface, the more severe the outcome. Similarly, any other features beneath the window, such as railings or fences, will affect the severity of the outcome.

In multi-storey residential buildings there is a need for increased safety precautions to upper-storey windows, because of the increased risk posed by the more severe harms resulting from the distance of the fall. In such buildings, and preferably from the second floor upwards, glazing below 1,100 mm from floor level should be guarded with a safety rail.

Collisions, entrapment, cuts, and strains – While there is always an inherent risk of entrapment by doors and windows, certain features can increase that risk. It is increased where a door or window is difficult to close, or where there is an over-powerful door closer (such that a small child may not be able to resist it). Doors and windows that pivot (rather than being hinged from one edge) can trap fingers or hands. Broken cords to vertically hung sliding sash windows will mean that the window cannot be operated without a risk of physical injury.

Doors opening into passages, small rooms (such as bathrooms), or onto stairs create a risk of a collision. Doors to wall-hung cupboards over worktops in kitchens can also be a risk. Doors or window opening lights should not project over pathways to obstruct the passage of those using the path.

As mentioned above, gaps, particularly in guarding to balconies, landings, and stairs, which are over 100 mm, may be attractive to small children who could become trapped.

Inadequate headroom presents a risk of collision. They can occur at door openings or at the top of and above stairs.

Door accidents resulting in injuries include doors shutting on, or trapping, part of a body. Most of such accidents appear to involve children aged under 10 years. There are also cutting and piercing injuries associated with glazing to both doors and windows. These can be limited or avoided by using safety glass in doors and windows in vulnerable locations (although not for bathrooms and WC compartments where privacy is required).

Strain and sprain injuries are those that result from poor ergonomics associated with the positioning and location of amenities, fittings, and equipment. The design and layout of dwellings has an effect on convenience of use, which can include awkward positioning of windows, difficult to operate window catches, inadequate functional space around facilities, and low worktop surfaces. As well as potential strains, inappropriate positioning of amenities and equipment can lead to other injuries (such as fall injuries), if an individual is forced to stretch or lean to reach a handle, catch, or switch. To avoid strain injuries wash hand basins, sinks, worktops, sanitary basins, baths, and showers should be at an appropriate height, and with sufficient free user space (functional space) to facilitate use without strain. Light switches should be sited conveniently for door openings and at each end of staircases and corridors, again at a reasonable height. Similarly, socket outlets should be conveniently sited. Door handles should be at a reasonable height and window catches should be readily accessible without strain. Cupboards and shelves should be sited where they can be easily reached, and without posing collision hazards.

Poisonings – There will be various potentially hazardous substances introduced into a dwelling by a household, including cleaning products and medicines. There should be lockable, child-proof cupboards to allow such substances to be stored safely.

Explosions and falling elements – Explosions in dwellings are usually associated with defective gas or hot water installations.

To avoid explosions from gas installations there should be appropriate, properly designed and installed, gas pressure regulators, meters, and pipework. The installation should be regularly serviced and tested to ensure there are no leaks or other defects. Gas appliances should be properly designed and installed, should satisfy relevant safety regulations, and should be fitted with automatic cut-off devices. The appliances and associated flues should be regularly serviced and maintained by a competent person.

Liquid petroleum gas is heavier than air, while natural gas is lighter. Where LPG is used, there should be adequate low-level ventilation or the means of ensuring any gas escaping can "drain" safely away. This is particularly important where the floor level is below the adjacent ground level. LPG containers and storage tanks should be secure and sited well away from possible sources of ignition.

Hot water systems should be correctly installed to meet the requirements of safety regulations. No hot water storage tank of more than 315 litres capacity should be connected directly to the mains water supply. For ventilated hot water systems, there should be an adequately sized vent pipe sufficient to allow steam to escape in case of thermostat failure. Unvented systems should be provided with both a non-self-resetting thermal cut-out and one or more temperature relief valves. These safety devices should be regularly tested.

Objects or elements of a dwelling can, by becoming detached or collapsing, cause a range of physical injuries. Externally, the risks can result from falling roof slates or tiles, eaves gutters, bricks or windows, wall cladding, and the collapse of walls. Internally, the risks can be the collapse of floors, ceilings, stairs, or a fixture (such as a light fitting or kitchen cabinet) becoming detached from the ceiling or wall. To avoid such risks, all elements of the structure of a dwelling should be properly maintained to ensure they remain safe and stable. The foundations and load-bearing external walls should be of sufficient strength to support the weight of the building, fittings, furnishings, and users. Any disrepair should not interfere with structural integrity, and any external cladding, rendering or similar finishing, and any coping should be securely fixed. All external balconies and walkways should be designed, constructed, and maintained so as to be capable of supporting their own weight and the imposed loads (such as plant pots) and persons.

The roof structure should be designed, constructed, and maintained so as to be strong enough to support the weight of the covering, be securely fixed, and to cope with wind- and weather-imposed loads (including snow).

Internally, floors should be designed, constructed, and maintained to be of sufficient strength to support their own weight and that of imposed loads including furniture, fixtures, fittings (including facilities such as baths and WC basins), and the occupiers. Stairs should be designed, constructed, and maintained to be of sufficient strength to support their own weight and that of imposed loads, including users and furniture likely to be carried up and down. Ceilings should be designed, constructed, fixed, and maintained to be strong enough to remain intact.

Internal walls should be designed, constructed, and maintained to be strong enough to support their own weight and any loads reasonably expected. Such loads could include upper floors and ceilings, shelves, pictures, light fittings, equipment, facilities, and fixtures. Door frames and openings should be properly fixed and maintained and capable of supporting the doors.

Shock and burn injuries

Electricity – Contact with metal or other conducting material that is "live" (i.e., connected to a supply of electricity) can result in an electric shock. The shock effect ranges from mild tingling sensations to disruption of the normal regular contractions of the heart or respiratory muscles, causing death. The effect depends on several factors, the main one being the voltage across the body.

As human tissue acts as a resistance, the heat generated may result in burns, usually occurring at the point of contact with the source of electricity.

While the majority of injuries are not severe, death can be an outcome. Fatalities can be caused by a defect or deficiency in the electrical wiring and other installations, and such defects may also be the source of ignition for a fire.

The potential danger of electrocution means that safety precautions should be incorporated. The precautions are directed to isolation and/or insulation. Potentially "live" components are covered with non-conducting material, all exposed metal parts of the installation are "earthed" (so that in the event of a deficiency any current will flow immediately to earth), and for installations at 230 volts AC or higher, a residual current device or residual current circuit breaker (RCD and RCCB) can provide additional safety and be incorporated into the consumer unit (the distribution boards immediately after the mains supply). RCDs and RCCBs are devices that detect some, but not all, deficiencies in the electrical system and rapidly switch off the supply.

Where there are insufficient and/or poorly located electric sockets there can be overloading of the circuit by use of extension cables or multi-socket adapters that can lead to fires (see below, p. 46), and trailing cables that provide a trip hazard.

As water is highly conductive, additional precautions are considered necessary in most countries for bathrooms, kitchens, and other areas where individuals could be in contact with both water and a source of electricity. This usually involves restrictions on electric socket outlets to 12 volt AC sockets, such as shaver sockets in bathrooms (although there are no restrictions in kitchens).

For tall and isolated buildings, a lightning protection system may be necessary where, depending on the geographical location, there is a risk of a lightning strike.

Hot surfaces, materials, and exposed flames – Contact with these sources can result in burns or scalds, and it is these sources that account for the great majority of non-fatal burn accidents. The severity of the burn or scald is dependent on its depth and the area covered. The depth of burn is dependent on the temperature of the hot object or liquid, the length of time of exposure, the time taken before corrective action is taken (such as the application of cold water). The length of time that hot material can be touched without damage to human tissue depends on the material, as well as the temperature.

Severe burns or scalds can result in permanent scarring. And, as well as obvious physical pain, many victims, and also parents of children involved, suffer acute psychological distress for many years. The relatively small body area (especially when hot liquids are involved), and the more sensitive nature of their skin, means that young children are particularly at risk of suffering severe injuries. Many of these victims suffer extensive full thickness burns and require plastic surgery, often for many years following the accident.

The kitchen is the site of many severe burn and scald injuries, particularly those to young children. These often involve accidents with cups and mugs of hot drinks, kettles, teapots, coffee pots, saucepans, cookers, and chip pans and deep fryers. Although behaviour is a major factor, the design and layout of the dwelling can contribute to these accidents, in particular, the design and layout of kitchens, the relationship between the kitchen and living/dining areas, the location of the cooker, and the design, siting, or adjustment of fixed heating appliances and radiators.

Appliances, such as gas and open fires and unfixed heaters, cause the most deaths from burns. These appliances and gas hobs may be involved in burns resulting from clothing catching alight.

Where cooking is carried out within a bedroom or living room, there can be an increased likelihood of an accident if the kitchen area is inadequately separated from the living or sleeping area. If there are insufficient numbers of electric socket outlets provided in the kitchen area, it can result in kettles, or other kitchen appliances, being used in non-kitchen areas, which may result in an increased risk of scalds.

In multi-occupied accommodation with shared kitchens remote from the unit of accommodation, then there may be an increased risk of burns and scalds associated with carrying hot drinks and food from the kitchen to the accommodation.

Uncontrolled fires – There are four main factors relevant to fire and fire safety. These are:

- The ignition and starting of a fire. All fitted appliances and equipment that present a possible source of ignition should be correctly and safely installed and maintained. The space for siting cookers should be safe, with no flammable materials immediately adjacent, and away from windows where curtains may be hung. All fixed heating appliances and systems should be properly designed, installed, and regularly serviced and maintained. Adequate means for space heating of the whole of the dwelling will discourage the need for, and use of, supplementary portable heaters. Facilities for drying clothes indoors during inclement weather will discourage placing clothing near to or on heaters. There should be sufficient and appropriately sited electric socket outlets to help reduce the need for extension leads and overloaded sockets. The electrical installation (distribution board, wiring, etc.) should be properly installed, maintained, and regularly checked and tested.
- Occupier behaviour is a major factor in relation to fires starting, including carelessness or misuse of equipment or appliances. Fires started by smokers' materials and matches are a major cause of fires and accidental deaths from dwelling fires. However, fires relating to cookers are attributable to misuse or carelessness by the occupier, and include chip pan fires, cooking left unattended, the use and siting of portable heating appliances (both electric and gas), and placing articles (clothing) too close to heaters or fires. There are some fires attributable to equipment deficiencies or the siting of the cooker (e.g., close to flammable materials). It also seems that a household with children is more likely to experience a fire compared with one without children; this may be because adults are distracted by children whilst cooking. Other than occupier behaviour, sources of ignition include cooking appliances, space heaters, and electrical distribution equipment.
- The design, materials, construction, and maintenance of the dwelling or building should limit the spread of fire, containing it within the room or dwelling of origin. This includes fire doors within the dwelling (usually the door to the kitchen) and, in the case of apartments, the design, etc.,

of the entrance door. Fires are generally confined to the room where the fire originated. Most fires start in the kitchen, fewer in bedrooms, and very few in circulation areas or other areas. In the case of apartments in a block, fires are (or should be) confined with the dwelling where the fire started.

- If (or when) a fire starts and begins to spread, occupiers need to be made aware of it. Properly working alarms, connected to smoke or heat detectors probably do more to save lives in the event of a fire than other factors. They provide early warning to the occupants, allowing them to escape before they are overcome by fumes or burned. The detectors and alarms should be appropriately sited, maintained and regularly tested. Battery-powered alarms have a relatively high failure rate (often a result of discharged or missing batteries), compared to that for mains-powered (hard-wired) alarms. (Note, there are alarms are available for those with hearing impairment.)

- There should be a means of escape from all parts of the dwelling or building, particularly for rooms and areas above the ground floor. In addition, there should be the primary means of fighting fire, such as a fire blanket close to cooking appliances and extinguishers. Based on UK statistics, the elderly and the very young (aged 4 and under) are most at risk. They both may have mobility problems that will affect the ability to, and speed of, escape. For any form of multi-occupied residential buildings, there should be adequate fire protection to the means of escape and between each unit of accommodation, appropriate fire detection and alarm system(s), and, as appropriate, emergency lighting, sprinkler systems. or other fire-fighting equipment

The most common cause of death from a fire is being overcome by gas or smoke. There are fewer deaths from the result of burns alone. As well as the physical outcomes, there are also mental and social outcomes associated with surviving fires. There is the trauma of the event, the loss of home and belongings, including the damage caused by smoke and from the fire-fighting, the stress associated with finding somewhere to live, and the stress related to dealing with authorities and agencies involved when addressing the consequences. Where there have been deaths or major injuries, there will be the stress from the loss of, or injuries/trauma to, relatives or friends.

It is not only victims that are affected by a serious fire. Close family of the victims are at an increased risk of deteriorating mental and physical health. Responders to fire disasters, such as fire-fighters, paramedics, and the police, are also at increased risk of mental health outcomes.[29]

There will also be social outcomes from the disturbance and disconnection from the local community and support networks.

Extreme events

These are events, often intermittent, that can pose severe danger to health and safety, and include earthquakes, flooding, and prolonged or extreme heat waves or cold events. One of the symptoms of climate change is extreme weather events and so these may become more frequent and of greater intensity.[30]

In regions where earthquakes can occur, buildings should be designed and constructed to be resilient.

Where flooding is likely, it is possible to design and construct dwellings with defences (such as removable barriers to door openings), and with internal wall finishes and floors that will be unaffected by water. However, the majority of defences will (or should be) outside the dwelling. These should include barriers, and appropriate seal mechanisms for drains and sewers to prevent foul sewage escaping. The effects of flooding and the loss of home through flooding cause severe trauma and have long-term health effects.[31]

To limit the effects of heat waves and cold events dwellings should be properly insulated, and ventilated. To limit overheating there should be means of shading (such as shutters, awnings, or blinds).[32] Where possible, precautionary measures should be passive, and air conditioning a last resource in order to avoid the need for energy use. To reduce the likelihood of prolonged exposure to low indoor temperatures, insulation and effective and efficient space heating should be provided.

Notes

1 Ormandy D (ed.) (2009) *Housing and Health in Europe: The WHO LARES Project*. Routledge, Oxon.
2 Carroll B, Morbey H, Balogh R, Araoz G (2008) Flooded Homes, Broken Bonds, the Meaning of Home, Psychological Processes and Their Impact on Psychological Health in a Disaster. *Health & Place* 15: 540–547; and Carroll B, Balogh R, Morbey H, Araoz G (2010) Health and Social Impacts of a Flood Disaster. *Disasters* 34(4): 1045–1063.
3 See the Whitehall studies, statistical database; Marmot M et al. (1978) Employment Grade and Coronary Heart Disease in British Civil Servants. *Journal of Epidemiology and Community Health* 32(4): 244–249. https://www.ucl.ac.uk/iehc/research/epidemiology-public-health/research/whitehall.
4 http://www.projecthelix.eu/fr. See also https://pubs.acs.org/doi/full/10.1021/acs.est.7b01097.
5 See http://www.healthyhousing.org.nz/research/past-research/housing-heating-and-health-study/; and http://www.healthyhousing.org.nz/research/past-research/housing-insulation-and-health-study/.
6 Ezratty V, Bonay M, Neukirch C, Orset-Guillossou G, Dehoux M, Koscielny S, Cabanes P-A, Lambrozo J, Aubier M (2007) Effect of Formaldehyde on Asthmatic Response to Inhaled Allergen Challenge. *Environmental Health Perspectives* 115(2): 210–214.
7 For example, Hamilton I et al. (2015) *BMJ Open* 5: e007298. doi:10.1136/bmjopen-2014-007298.

8 A source of recent research reports is available at RHE Global International Housing and Health Research Bulletin. https://www.housinghealth.com.

9 BD2518–availableat:http://webarchive.nationalarchives.gov.uk/20120919135603/; http://www.communities.gov.uk/publications/planningandbuilding/review healthsafety.

10 Including: WHO Housing and Health Guidelines (2018) World Health Organization, Geneva. NCHH and APHA (2014), https://nchh.org/resource-library/national-healthy-housing-standard.pdf.

11 Ormandy D, Ezratty V (2012) Health and Thermal Comfort: From WHO Guidance to Housing Strategies. *Energy Policy* 49: 116–121.

12 Public Health England (2014) *Minimum Home Temperature Thresholds for Health in Winter: A Systematic Literature Review*. Public Health England, London.

13 Public Health England (2014) *Minimum Home Temperature Thresholds for Health in Winter: A Systematic Literature Review*. Public Health England, London.

14 Dannemiller KC, Gent JF, Leaderer BP, Peccia J (2016) Influence of Housing Characteristics on Bacterial and Fungal Communities in Homes of Asthmatic Children. *Indoor Air* 26: 179–192. doi:10.1111/ina.12205.

15 Dampness and Mould (2009) *WHO Guidelines for Indoor Air Quality*. http://www.euro.who.int/__data/assets/pdf_file/0017/43325/E92645.pdf?ua=1.

16 Royal College of Physicians (2016) *Every Breath We Take; The Lifelong Impact of Air Pollution*. Report of a Working Party. London. https://www.rcplondon.ac.uk/projects/outputs/every-breath-we-take-lifelong-impact-air-pollution.

17 Cheng M et al. (2016) Factors Controlling Volatile Organic Compounds in Dwellings in Melbourne, Australia. *Indoor Air* 26: 219–230. doi: 10.1111/ina.12201.

18 https://nchh.org/information-and-evidence/learn-about-healthy-housing/lead/#Health_Impacts

19 https://www.ukradon.org/information/ukmaps. There are similar maps for other countries.

20 Office of the Deputy Prime Minister (ODPM) (2004) *The Impact of Overcrowding on Health and Education: A Review of the Evidence and Literature*. Office of the Deputy Prime Minister, London.

21 https://www.who.int/sustainable-development/publications/housing-health-guidelines/en/.

22 Office of National Statistics (ONS) (2017) *Public Perceptions of Crime in England and Wales*, year ending March 2016. Office of National Statistics, London.

23 https://www.bre.co.uk/filelibrary/Briefing%20papers/Lighting-and-health-infographic2.pdf.

24 http://www.euro.who.int/en/health-topics/environment-and-health/noise/environmental-noise-guidelines-for-the-european-region.

25 WHO (2018) *Environmental Noise Guidelines for the European Region*, Copenhagen, Denmark. http://www.euro.who.int/__data/assets/pdf_file/0008/383921/noise-guidelines-eng.pdf?ua=1.

26 Bonnefoy X, Kampen H, Sweeney K (2008) *Public Health Significance of Urban Pests WHO Regional Office for Europe*, Copenhagen, Denmark.

27 Battersby SA (2015) Rodents as Carriers of Disease. In Buckle AP, Smith RH (eds), *Rodent Pests and Their Control*, 2nd edn. CABI, Wallingford, Oxon, pp. 81–100.

28 https://www.cdc.gov/homeandrecreationalsafety/falls/adultfalls.html.
29 The impact of fire on residents can be similar to that caused by flooding (see note 27).
30 Murray V (2017) Extreme Events – The Sendai Framework for Disaster Reduction. In Battersby SA (ed.), *Clay's Handbook of Environmental Health,* 21st edn. Routledge, Oxon, pp. 976–992.
31 Carroll B, Morbey H, Balogh R, and Araoz G (2008) Flooded Homes, Broken Bonds, the Meaning of Home, Psychological Processes and Their Impact on Psychological Health in a Disaster. *Health & Place* 15: 540–547; and Carroll B, Balogh R, Morbey H, Araoz G (2010) Health and Social Impacts of a Flood Disaster. *Disasters* 34(4): 1045–1063.
32 https://www.bre.co.uk/filelibrary/Briefing%20papers/116885-Overheating-Guidance-v3.pdf.

Part 5

Standards and guidelines

These form the basis of any assessment of dwellings and of the mechanisms for ensuring compliance. How they are applied depends on the legal and administrative structure of the particular jurisdiction. The legal and administrative environment can differ considerably, and what may be an appropriate and effective option in one jurisdiction may prove to be ineffective in another. Here, the various types of standards and guidelines area discussed.

Formulation of standards

Specific or quantitative standards specify what should or should not be present, or specify something that is measurable. These have the advantage of being clear and easily understandable and can be applied by relatively untrained staff. In housing terms they are most suited for controlling standards in new buildings, where a specific requirement or measurement can be met – the height of a window sill above the floor, or the thermal transmittance of a wall or roof. Disadvantages include that, being building focused, means that whether a defect is seen as serious is based on the extent or cost of the remedial works necessary rather than the potential for harm from the defect. While such an approach is particularly useful in controlling standards in new (yet to be built) housing, it is not so for existing buildings.

A vague or qualitative standard is one that states what should be taken into consideration when designing or assessing conditions or situations in order to mitigate or avoid threats to health and/or safety.

The majority of the housing stock will have been constructed to satisfy previous standards, or perhaps no standards for certain elements, so some elements and features may not satisfy current requirements for new buildings. The most appropriate approach for existing buildings is through qualitative "standards". These are ones that use relatively vague terminology, such as "as safe as reasonably practicable", which are applied by a qualified assessor able to use their informed professional judgement. In housing terms they need to be able to assess the potential threat to health and/or safety posed by a defect or deficiency to the dwelling – the effect of steep stairs with no handrails, or of an unexpected small change in floor level (a "trip step"). Rather than being building focused, qualitative standards are more likely to be human focused, so that the seriousness of the defect or deficiency is judged in terms of the severity of the health/safety outcome. This approach is particularly suitable for existing dwellings. However, because of the need for qualified personnel to apply them, they can be expensive to implement and may not be readily understandable by everyone; but there will be less need to update the standards as the onus is on the qualified personnel to keep up-to-date on the potential threat from conditions.

Any standards or guidelines relating to housing should be geared towards whether the structure and facilities enable the premises to be used as a dwelling, and do not interfere with the occupier(s) establishing a home.

Minimum, maximum, and target standards

A minimum standard is one that states what is the least necessary, such as the provision of a supply of water, the provision of at least one sanitary facility, or at least one bath or shower with a supply of hot and cold water.

A maximum standard states a level that cannot be exceeded and usually applies to pollutants, such as the maximum parts per million of a particular air pollutant a person can be safely exposed to over a given period.

Minimum and maximum standards are often used to control new (yet to be built) dwellings. However, while these can be seen as a simple and straightforward way to protect health and safety, they can become the norm, i.e., what is the incentive to go beyond what is necessary to meet the standard? If the method of updating standards is slow or otherwise inadequate, the standard itself may lag behind research on housing and health and therefore fail to safeguard health as intended.

Standards can also be used to trigger intervention. If something fails to meet the standard, then a regulatory body can or must intervene to require remedial action. There are options on the nature and extent of the action that could be required, one of which is to require that a higher or target standard is met. This last option should mean that it is unlikely that there will be deterioration for some time.

Such standards may also be used to trigger financial assistance (grants or loans) towards the cost of taking the necessary action.

A target standard sets what should be achieved after intervention and should be higher than that which triggered the intervention. The aim being that the result should extend the life of the dwelling for 15 or more years, so avoiding the need for intervention for some time.

Administrative standards or guidelines

These have no legal force but can be used to control the performance or production of an agency. For example, a local authority or agency responsible for the provision of social (public) housing may be required to meet certain standards before the central department releases finance or permits the expenditure. The formulation (phrasing) of such standards can be descriptive, or in the form of guidance to be taken into account. The "sanction" for these standards is the denial of the release the funding or the authorisation of expenditure. One such standard in England is the Decent Homes Standard.[1]

Setting standards and drafting guidelines

There are various stages involved in setting standards and drafting guidelines.

First, a potential threat to health or safety is identified, usually based on or supported by evidence to establish the threat. Depending on the type of threat and how to avoid, remove, or minimise that threat will determine the setting of standards or guidance.

For example, an approach appropriate for threats related to indoor air pollutants, such as carbon monoxide, and sulphur dioxide, is to specify or recommend maximum safe exposure levels. The maximum limits could be those recommended by the World Health Organization (WHO) guidelines for indoor air quality and WHO guidelines for ambient air quality; similarly, the WHO recommendations for minimum and maximum thresholds for indoor temperatures could be adopted. A similar approach could be adopted for noise, again using the WHO guidelines for noise to set the maximum safe limits. However, in the case of air quality and temperature in particular, measuring whether or not the thresholds have been breached can be problematic (on which see Part 6, p. 57).

Where the lack of something would create a threat, such as a lack of a supply of water, or the lack of sanitary/hygiene facilities, the approach would be to specify what should be present. Conversely, where the presence of something, such as dampness and/or mould growth, creates a threat, the approach would mean stating that it should not be present. Again there may be complications in stating the extent and position of the threat (e.g., the area covered and the location – in a hall or passageway would pose less of a threat than in a bedroom).

The English and Welsh Housing Health and Safety Rating System (HHSRS)

The HHSRS is unique, in that it is not a standard, but a health-based risk assessment approach to the evaluation of housing conditions, shifting the focus from defects and deficiencies in the structure and facilities to the potential threat to health and/or safety attributable to the conditions.[2]

The system was developed by the University of Warwick Law School, with support from the UK Building Research Establishment (BRE) and the London School of Hygiene and Tropical Medicine (LSHTM).[3] The HHSRS was incorporated into legislation as the prescribed statutory method for assessing housing conditions in England and Wales in 2006, was included in the English House Condition Survey (EHCS; now the English Housing Survey (EHS)), and was adopted as the first criterion of the Decent Homes Standard.[4]

In 2010, the HHSRS was adopted by the US Department for Housing and Urban Development (unchanged, except for the name, the Healthy Homes Rating System) and used as a condition for an award of a grant for projects dealing with housing conditions.[5] The HHSRS was also used to inform the development of the New Zealand Healthy Housing Index (HHI).

Based on an extensive literature review, 29 potential housing hazards were identified, each to a greater or lesser extent attributable to the condition of the dwelling. These hazards were then linked to health outcomes.

While a deficiency may have implications in building and aesthetic terms, for the purposes of the HHSRS its only relevance is whether the effect from that deficiency has the potential to cause harm, i.e., results in a hazard.

A single deficiency may contribute to more than one hazard. For example, disrepair to a ceiling, dependent upon the nature and extent of that disrepair, could lead to the following hazards:

- excessive cold (through increased heat loss);
- fire (by allowing fire and smoke to spread to other parts of the dwelling);
- lead (from old paint);
- infections from other sources (by providing means of access and harbourage for pests);complete collapse; and
- noise (because of an increase in noise penetration between rooms).

The opposite is also possible, where several deficiencies contribute to the same hazard. Disrepair to a ceiling, an ill-fitting door, and the lack of a smoke detector may all contribute to a fire hazard, as each could lead to smoke and flames spreading through the dwelling.

The UK Decent Home Standard

This is an administrative standard and guidance used in both the public and private sectors.[6] A dwelling is deemed to have failed to meet this standard if:

- it contains one or more totally unacceptable HHSRS hazards;
- it is not in a reasonable state of repair;
- the facilities and services are more than 20 years old;
- it does not provide a reasonable degree of thermal comfort (having regard to the HHSRS hazard of excess cold).

New Zealand Healthy Housing Index and the Rental Warrant of Fitness (RWoF)

Both of these follow a checklist approach and are not so much a standard, but more a means of assessment with a list of the matters to be taken into account. The HHI is a tool to understand the link between housing and health at both a community and individual household unit level.[7] It is important to note that the HHI focuses on the structure and condition of the house (it does not take account of occupier behaviour). It is intended to identify the potential for occupants to have a safe and healthy interaction with their dwelling. In this sense the HHI is a snapshot within a broader picture of what it means to have a healthy home. The RWoF is based on the HHI and is being proposed as a means to certify the suitability of a dwelling for letting.[8]

French Grid for the Evaluation of Dwellings Likely to be Declared Unhealthy

The "Grille d'Evaluation de l'Etat des Immeubles Susceptibles d'Etre Declares Insalubres" is a guide and set of assessment sheets. It was developed by those who were assessing and dealing with housing conditions, and it is practical rather than a scientific work. The guide is a collection of analyses, comments, and observations from technicians to advise other technicians.

The assessment grid includes 35 criteria for buildings and 29 for dwellings. It contains 21 sheets corresponding to the total of 18 criteria for evaluating the state of the building, and 13 criteria for assessing the condition of individual dwellings. This approach resulted from the choices made by the working group to prioritise certain topics considered more important.

The result of the assessment is given on a scale of 0 to 1, each element being given a weighting reflecting its perceived importance and condition. The condition of individual elements requires a judgement as to whether it is satisfactory, requires some minor work, requires major work, or is dangerous/needs replacing.

World Health Organization Housing and Health Guidelines

WHO states that these guidelines bring together the most recent evidence to provide practical recommendations to reduce the health burden due to unsafe and substandard housing.[9] Based on commissioned systematic reviews, the guidelines provide recommendations relevant to inadequate living space (crowding), low and high indoor temperatures, injury hazards in the home, and accessibility of housing for people with functional impairments. In addition, the guidelines identify and summarise existing WHO guidelines and recommendations related to housing, with respect to water quality, air quality, neighbourhood noise, asbestos, lead, tobacco smoke, and radon. The guidelines take a comprehensive, intersectoral perspective on the issue of housing and health and highlight co-benefits of interventions addressing several risk factors at the same time.

US National Healthy Housing Standard

This standard was produced jointly by the National Center for Healthy Housing and the American Public Health Association.[10] It is an evidence-based standard intended to be a tool to reconnect the housing and public health sectors, and a standard for those in the position of improving housing conditions. It was drawn from the latest (2014) and best thinking in the fields of environmental public health, safety, building science, engineering, and indoor environment quality. As well as individual requirements, it gives so-called "stretch provision", i.e., additional matters over and above the basic requirements.

Notes

1 https://assets.publishing.service.gov.uk/government/uploads/system/uploads/attachment_data/file/7812/138355.pdf.
2 Office of the Deputy Prime Minister (ODPM) (2006) *Housing Health and Safety Rating System: Operating Guidance*. Office of the Deputy Prime Minister, London.   https://assets.publishing.service.gov.uk/government/uploads/system/uploads/attachment_data/file/15810/142631.pdf.
3 Reports on the development are available at http://sabattersby.co.uk/hhsrs.html.
4 https://assets.publishing.service.gov.uk/government/uploads/system/uploads/attachment_data/file/7812/138355.pdf.
5 https://www.hud.gov/program_offices/healthy_homes/hhrs.
6 https://assets.publishing.service.gov.uk/government/uploads/system/uploads/attachment_data/file/7812/138355.pdf.
7 http://www.healthyhousing.org.nz/research/past-research/healthy-housing-index/.
8 http://www.healthyhousing.org.nz/wp-content/uploads/2016/09/WOF_Assessment_Criteria_and_Methodology__Version-3.0.pdf.
9 https://www.who.int/sustainable-development/publications/housing-health-guidelines/en/.
10 https://nchh.org/tools-and-data/housing-code-tools/national-healthy-housing-standard/.

Part 6

Inspections and assessments

Before any standards or guidelines can be applied, there needs to be an assessment of the dwelling, i.e., an inspection. The inspection should be based on the use of the premises as a dwelling, and the potential effect(s) of any conditions, defects, or deficiencies that interfere with that use. In housing and public health terms, the principle behind the assessment of a dwelling is to determine whether the basic physiological and psychological requirements for human and family life and comfort are satisfied. It also means that standards or guidelines should be geared to determining whether the structure and facilities do not interfere with the occupier(s) establishing a home.

The general aim in carrying out any inspection of an individual dwelling is to collect information on the design, construction, layout, state of repair, facilities, and deficiencies. Only after a whole dwelling inspection is it possible for the condition to be compared to any housing standard or guideline, and a decision made as to whether the dwelling meets the standard or where it "sits" on a scale.

As discussed in Part 5, there are various potential approaches to inspections, these include:

- An inspection by a qualified person using informed professional judgment. In this case the person carrying out the inspection would be required to consider whether a facility, element, or component satisfies its function.
- A checklist approach. For this a list of what should or should not be present. For a more sophisticated approach, the state and condition should also be assessed (i.e., satisfactory, minor repairs, major repairs, or replace). In its simplest form, a checklist approach can be applied by a relatively untrained person, who is required merely to observe the presence or absence of a facility, element, or component. The more sophisticated approach would need a person with some knowledge of

the use/purpose of the facility, element, or component, so as to be able to give an opinion on whether it satisfies that function, and comment on its condition.

Assessors

In the UK, housing inspections and assessments are usually made by an environmental health practitioner (EHP), a qualified generalist. EHPs are responsible for the protection of public health, including administering and enforcing related legislation on behalf of the local authority. They also provide support to minimise health and safety hazards. EHPs are multiskilled and deal with several environmental matters; they are qualified, usually to degree level, and follow continuing professional development. For particular problems, the EHP will call in a specialist for advice and proposals for the appropriate remedy – for example this could be for the assessment of energy efficiency, electrical issues, or gas issues.

In some other countries, individual specialists are called in to assess a particular problem.

Generally, the inspection and assessment are of the dwelling (the structure and facilities), but there are a few examples of assessments of the dwelling together with the occupier behaviour – the dwelling as occupied – recognising that there is an interaction between the occupying household and the dwelling. The following are three examples where both the dwelling and the household behaviour are assessed. These are usually commissioned by a medical doctor or hospital dealing with a patient suffering from a respiratory condition (such as asthma).

The Children's Mercy Hospital in Kansas, US, runs courses on assessments focusing on both the condition of the dwelling and the behaviour and activities of the household. The inspection and assessment are triggered by the hospital as a reaction to a patient with respiratory conditions. The courses offered cover what is considered the essential knowledge of eight healthy home principles and how to apply these when assessing health and safety risks in dwellings. The assessment includes taking an environmental history from the clients, and providing in-home education. The assessors' course also provides basic building science, visual assessment techniques, identifying and prioritising health and safety concerns, basic environmental measurement and sampling, and risk communication and assessment reporting.[1]

In 2000, Brussels, Belgium, created Green Ambulances ("ambulances vertes", Cellule Régionale d'Intervention en Pollution Intérieure, CRIPI).[2] A medical doctor can commission an inspection and assessment covering both the dwelling and occupier behaviour. The response is a team including analysts and a social nurse, and the assessment involves taking chemical and biological samples and completing a questionnaire. Results of the analyses and from the interview are given to the doctor, and specific advice is given. There is a follow-up to check the health of the patient.

In France in 2001, a diploma course was set up to train Medical Indoor Environment Counsellors (CMEI).[3] Again, a medical doctor can ask for an inspection of a dwelling by a CMEI whenever he or she suspect that the health of a patient could be caused or exacerbated by the condition of their dwelling and/or the occupier's activities, for example in case of asthma. The CMEI will visit every room of the dwelling and take account of the heating, ventilation, furniture, coverings, carpets, plants, etc. The CMEI will also interview the occupier about recent repairs, and about their behaviour and activities. Advice will be given about organisations they could seek advice from, such as Agence Nationale d'Amélioration de l'Habitat (ANAH), Association Départementale d'Information sur le Logement (ADIL), or Conseil d'Architecture d'Urbanisme et d'Environnement (CAUE). The CMEI will also prepare a report, with recommendations and actions to be taken, which is sent to the medical doctor and the patient. CMEIs are available in every region of France and the visit is free when it is done by a CMEI not in private practice (see: http://www.cmei-france.fr/index.php?section=1-accueil-du-site-des-cmei).

Alternatives to standards

Standards usually have a binary, pass/fail approach. An alternative is to provide a form of grading. An example of this is the Housing Health and Safety Rating System (HHSRS) in England and Wales. This provides a risk-based methodology for assessing the immediacy and severity of a threat to health and/or safety attributable to the condition.

By grading the immediacy and severity of any threat, the HHSRS allows for the targeting of action, both within the dwelling itself, giving priority to the most serious hazard(s), and between two or more dwellings, targeting the most hazardous dwelling. As the HHSRS does not draw a pass/fail line it also encourages the improvement of any dwelling.

Assessment of dwellings

Assessing a dwelling should be based on determining whether it satisfies the various functions and requirements as set out in Part 4. Generally, this is straightforward, but can be complex in some areas, some of which are discussed and expanded on below.

Assessing thermal comfort

Although there is guidance giving a range of recommended indoor temperatures there are practical problems associated with physically taking temperature readings. The indoor temperature will depend on various factors including the time of year, the time of day, and household activities. In addition, there are both horizontal and vertical temperature gradients within rooms and within dwellings.

Data on external ambient temperatures is widely available. However, there is little data on indoor temperature or factors in the immediate outdoor environment that will influence the indoor temperature (e.g., urban heat island effects) there has been limited investigation into the relationship between the external and internal temperatures. A study in Boston suggested that the association between indoor and outdoor temperatures was good at warm temperatures, but weak at cooler temperatures, while another study found a close relationship between outdoor temperatures and indoor temperatures in poorly thermally insulated dwellings.

Recognising that there is a relationship between indoor and outdoor temperatures, the concept of heating degree days and cooling degree days has been developed in the US to assess the influence of temperature change on energy demand by measuring the difference between outdoor temperatures and the temperatures that people generally find comfortable indoors.[4]

A strict protocol can be adopted to predict thermal comfort, taking account various factors that could affect it. However, this is not possible for housing surveys, where strict protocols have not been devised and adopted. Several alternatives have been used, including modelling or predicting thermal comfort, but these are not really appropriate for the housing situation.

The best-known method for predicting thermal comfort has been developed by the American Society for Heating, Refrigerating, Air-conditioning Engineers (ASHRAE) that uses a thermal sensation scale based on the laboratory work.[5] Similar methods have been proposed by the International Organization for Standardization (IOS),[6] and the Chartered Institute of Building Services Engineers (CIBSE).[7]

The ASHRAE approach uses a formula to calculate the Predicted Percent Dissatisfied (the PPD), i.e., the percentage of people who would be uncomfortable with a particular thermal environment. This includes physiological variables and the estimated thermal load on the body to produce the Predicted Mean Vote (PMV) index. The PMV index aims to give the mean response of a large group of people to a thermal sensation scale ranging from "hot" to "cold", with numbers used to represent the responses ("+3" through to "–3"). Where the PMV index moves away from zero it predicts that there will be a higher percentage of occupants dissatisfied as they will be either too warm/hot or too cool/cold. The rationale is directed at minimising the percentage dissatisfied, although it is stated that even with a PMV equal to 0, about 5 per cent of the people will be dissatisfied. This prediction approach, including the adaptive model, is inappropriate for the housing environment. As stated by ASHRAE, it is intended to predict the PPD of a "large group of people", and dwellings are rarely occupied by a large group. Also, 5 per cent of a household, even a large one, is meaningless.

Another possibility is to calculate whether a dwelling provides adequate protection against thermal discomfort using something like the approach for Energy Performance Certificates (EPCs), as used in the UK. The equivalent of EPCs are now required throughout Europe whenever a dwelling is sold or offered for rent, and they are intended to estimate the energy use of dwellings.[8] However, while the certificates are visually similar, the background calculations are very different.[9]

In addition, there is no direct relationship with the health of occupiers; an EPC (or its equivalent) can give an indication of whether the characteristics of a dwelling are such that an average household could avoid exposure to high or low temperatures. As the EPC calculation is based on energy efficiency, it could be used as a proxy to predict that an average household (one on a reasonable income) could maintain temperatures within the thermal comfort range.

Recognising that prediction methodologies are inappropriate for housing, and the limitations of using the EPC approach, most studies have used occupant perception. Using perception as a proxy for thermal comfort has several advantages, including taking into account the differences and period of exposure between members of the same household (age, gender, and susceptibility). For example, it seems that females have a greater need for individual temperature control. This perception approach was used in the WHO LARES project to assess the relationship between thermal discomfort and health in 8,519 individuals living in 3,382 dwellings in 8 European cities.[10] However, using residents' perceptions has some limitations, particularly for those age groups more susceptible to health threats from thermal discomfort. The very young may not be able to express their discomfort, and the elderly may be less sensitive, and may not recognise their exposure to excessive low or high temperatures. Furthermore, those over the age of 65 were found to be more satisfied with worse conditions, such as the lack of central heating, in the WHO LARES study.

In 2012/2013, a national survey of French housing and energy was carried out – Performance de l'Habitat, Equipements, Besoins et Usages de l'énergie (PHEBUS).[11] As well as interviews with residents of 5,405 dwellings representative of the 28 million principal residences in metropolitan (mainland) France, data were gathered on the energy equipment and the energy use behaviour. It also included information on a subsample of 2,389 dwellings to give a picture of the theoretical energy performance of these principal residences. PHEBUS also includes details of subjective satisfaction with the heating.

While there have been many studies into the health impact of heat, the majority of work on housing has concentrated on exposure to low temperatures. It has also been suggested that most people are more concerned about low temperatures than they are about high temperatures, perceiving low temperatures as potentially more threatening. This may interfere with responses to questions on thermal comfort, as may the time of year when the questions are posed.

As an additional point, in some countries, particularly those with a continental climate, "fuel poverty" (energy precariousness) includes not being able to afford sufficient energy not only to avoid low indoor temperatures, but also to avoid high indoor temperatures (i.e., avoiding temperatures below 18 °C and above 24 °C, the temperature range recommended by the World Health Organization).[12] Ideally, natural cooling is preferable to air conditioning, but it should be noted that the inability to afford sufficient energy for air conditioning, and the lack of protection from heat gain during heat waves, can have a serious effect on health, particularly of susceptible groups.

Assessing the thermal indoor temperature involves taking account of the functions of both the dwelling characteristics and the occupying household. For the dwelling characteristics, energy efficiency and the effectiveness of the heating system should be considered. For overheating, it is the provision for ventilation (particularly night-time ventilation) and provision for shading (including shutters and blinds). While air conditioning is a solution to overheating, it requires energy and maintenance.

As discussed above, a simple measurement of indoor temperature is inappropriate and impractical. The assessment should take account of the adequacy of the heating, insulation, and ventilation, although this may involve modelling, using a methodology such as the UK Standard Assessment Procedure (SAP) to determine the energy performance of the dwelling.[13] It should also include other factors that affect the indoor temperature, such as dampness, or disrepair to the structure, the thermal capacity of the structure, the orientation (particularly the glazed areas) and exposure of the dwelling.

Some of the dwelling characteristics and factors to be considered include:[14]

- Whether the thermal insulation of the external fabric (walls, floor, and roof) is adequate, both to retain heat and to prevent overheating (particularly in top floor rooms or dwellings).

- That the design and construction of the dwelling takes account of the orientation and exposure of the dwelling to limit excessive heat loss or heat gain. They should take account of the glazed areas.
- The presence of dampness that may reduce the effectiveness of the thermal insulating material and/or the structure.
- The provision for space heating should be appropriate for the type of construction, and the material used, and adequate for the size and layout of the dwelling. The system should have appropriate and usable means of control.

Assessing dampness and associated mould growth

Moisture is naturally present in many of the materials used in buildings and, provided that the amount remains within certain limits (dependent on the particular material), it need not present a problem. It is when the amount of moisture exceeds that limit that problems can occur – it is this excess moisture content that is usually referred to as "dampness". There are different forms of dampness, ranging from penetrating dampness due to a failure of the structure to prevent rain penetration through to condensation. It is sometimes a misconception that condensation is not "dampness" and housing providers and managers can be too quick to deny responsibility or blame the occupiers or their "lifestyle". Whatever the form of dampness, identification of the cause is crucial.[15]

Moisture (in the form of a gas – water vapour) is present in the air, and, provided it is within certain limits, it will not cause problems. The amount of water vapour a given amount of air can hold depends on the temperature of the air – the warmer the air, the greater the amount of water vapour it is capable of holding. The "relative humidity" is the amount of water vapour the air is holding compared with the amount it is capable of holding at that temperature, expressed as a percentage. The relative humidity within a dwelling should be between 30 per cent and 70 per cent, but there are both temperature and therefore relative humidity gradients within rooms and within the dwelling. Where the relative humidity persistently exceeds 70 per cent, problems will begin to occur; and when the relative humidity reaches 100 per cent, the air will give up some of the moisture in the form of condensation.

As stated above, many materials have a natural (and safe) moisture content. House bricks will have a moisture content of 1.5 per cent to 2.5 per cent, plaster around 1 per cent, and timber around 11 per cent. Where the moisture content exceeds these levels, there will be problems of "dampness".

Condensation dampness is frequently linked to energy inefficiency – where warm moisture-laden air comes into contact with a cold surface and gives up some of its moisture in the form of condensation. Moisture production is produced by the natural, domestic, and biological activities of the household. Relatively low levels of moisture are generated through breathing and are spread out over the 24 hours. Peaks of higher levels are generated from cooking, clothes drying, and bathing (or showering). Vapour pressure equalises humidity throughout a dwelling, so that high levels in one area will have an impact on the relative humidity in other parts.

Both house dust mites and moulds flourish in damp or humid conditions, and their growth is also influenced by temperature. Where the humidity exceeds 70 per cent, and temperatures are around 20 °C plus, there will be increases in mite populations and in mould growth.

The detritus from house dust mites and spores generated by mould growth are potent airborne allergens. The exposure to high concentrations of these allergens over a prolonged period can result in the sensitisation of atopic individuals (those with a predetermined genetic tendency to sensitisation) and may also sensitise non-atopic individuals. Once sensitised, relatively low concentrations of the airborne allergen can trigger allergic symptoms in an individual, such as rhinitis, conjunctivitis, eczema, cough, and wheeze. Repeated exposure can also lead to asthma attacks.

Spores from a wide range of fungi can cause allergic reactions, and this includes timber-attacking fungi. Fungal infection, while uncommon, can affect vulnerable individuals such as those on immuno-suppressant drugs. In addition, toxins from some moulds (mycotoxins) can cause nausea and diarrhoea, can suppress the immune system, and have been implicated in cancers.

The potential mental and social health impacts of dampness and mould should not be underestimated. Damage to decoration from mould- or damp-staining, and the smells associated with damp and mould can cause depression and anxiety. Also, feelings of shame and embarrassment can lead to social isolation (avoiding inviting friends and relatives into the dwelling). Recognising the public health importance of dampness and mould, the World Health Organization have issued guidelines.[16]

There is a wide range of variables that influence the likelihood of dampness, and this means that the assessment is one of judgement rather than measurement. Assessment should take account of the design, construction, condition, and state of repair of the dwelling. The location, extent, and duration of any dampness identified will affect the dust mite populations and mould growth, and the consequent potential for harm.

The immediate local climate and exposure will also affect the susceptibility of a dwelling. Localities with high levels of rainfall will increase the possibility of penetrating dampness. Similarly, the altitude and wind exposure of the dwelling will affect the thermal efficiency and associated condensation/high relative humidity (see "Indoor thermal environment" above, p. 26). This means that the assessment of a dwelling should take account of the prevailing weather conditions through all seasons, not just at a particular time. Penetrating and rising dampness may be less prevalent during dry weather, and condensation is less likely during warm periods of the year.

The size of a dwelling and the size of the occupying household are also factors – the size should be appropriate for the size of the occupying household; a mismatch such as a large household occupying a small dwelling (crowding) can lead to a problem of condensation (and other health problems).

The location of the damp and particularly mould growth is also important. The threat from damp in a bedroom, where individuals will spend around 8 hours out of 24, is greater than in other areas, such as bathrooms (where little time is spent). As well as mould on surfaces, etc., damp mattresses will support larger dust mite populations than other furniture and furnishings.

The cause of the dampness is also relevant to the assessment. While condensation is a symptom of high humidity, rising and penetrating dampness are related to the state of repair of the structure.

For dwellings where rooms are occupied for living (perhaps including cooking) and sleeping, such as bedsits and small flats in multi-occupied buildings, then the presence of dampness may be more significant as moisture generation will be high, and occupants can be expected to spend a greater proportion of their time exposed.

The various matters to be considered include:

- the energy efficiency of the dwelling (on which see "Assessing thermal comfort" above, p. 60);
- the general background ventilation, including the provision of means for the extraction of moisture-laden air during cooking, bathing, or showering, and the provision of clothes drying facilities;
- the presence of effective damp-proofing to walls, floors, and roofs and around window and door openings.

Assessing indoor air quality

(NB: This section focuses on non-biological pollutants. For potential biological pollutants, such as mould spores and dust mites (see "Assessing dampness and associated mould growth", etc., above, p. 63). For other possible biological pollutants, such as viruses, etc., see "Assessing space" below, p. 69.)

As discussed in Part 4, there are several possible health-threatening non-biological pollutants, including:

a)   asbestos
b)   biocides
c)   carbon monoxide and fuel combustion products
d)   uncombusted fuel gas
e)   volatile organic compounds (VOCs)
f)   lead dust
g)   radiation (in the form of radon gas).

Measuring or sampling for these is problematic and will depend on several factors, such as the household characteristics of the household, the household activities, the time of day, the time of year, the location, and the age of the dwelling. It will also depend on the period over which any measuring is carried out. Also important is the provision for ventilation – the means of removing possible pollutants from the interior and replacing them with fresh air. This can be problematic in urban areas, as ventilation can also introduce polluted air from outside.

As measuring or sampling is problematic, any assessment should be based on indications of the possibility of the pollutants.

Asbestos – In the past, this material was incorporated in a range of building products, including roofing, cladding, thermal and acoustic insulation, and fire-resistant internal panelling. This means that in many pre-1980, traditionally built houses and flats some products and materials containing asbestos (mostly chrysotile) may be present. However, airborne fibre levels in such dwellings are unlikely to exceed ambient background levels, as, in general, such materials were not in locations that were not likely to be disturbed (within the structure, rather than exposed surfaces).

Non-traditionally or "system-built" blocks of apartments, particularly those constructed between 1945 and 1980 may contain chrysotile and amphibole asbestos products, such as sprayed coatings and partitioning in site positions vulnerable to damage and disturbance. It is these dwellings where there could be a release of airborne fibres and the consequential risks to health.

Indoor air concentrations of asbestos fibres where the materials containing asbestos fibres is present but in good condition or sealed, present minimal risk to health. However, if the asbestos is damaged or deteriorating, then action should be taken to repair, seal, enclose, or remove the asbestos. Removal of asbestos requires specialist contractors, and will necessitate the temporary rehousing of residents.

The various matters to be considered include:

- The date of construction of the dwelling;
- the suspected presence of unsealed asbestos-based materials, particularly in potentially vulnerable locations;
- damage or disrepair to asbestos-based materials.

Biocides – The use of biocides should be avoided if possible. But where the use is necessary (such as treating timber before installation), then it should be in accordance with the instructions and the various legal controls. During and after use, thorough ventilation will be necessary to allow for fume dispersal, and this should include time before the dwelling is reoccupied.

Biocides are sometimes used to remove mould or fungal growth (which is likely to be a consequence of dampness) or to treat insect attack, the underlying cause of which will often have arisen as a consequence of dampness. As well as treatment of the mould or fungal growth, there should be action to deal with the underlying cause, that is, treatment of the source of dampness.

Carbon monoxide, etc. – As stated in Part 4, there are several potentially harmful products associated with the combustion (or more likely, the incomplete combustion) of carbon-based fuels, including gas, oil, and solid fuels used for heating and cooking. These include carbon monoxide (CO), oxides of nitrogen (NOx), sulphur dioxide (SO_2), and smoke (containing particulates including $PM_{2.5}$ and PM_{10}).

Wherever there are any gas, oil, or solid fuel burning appliances (including cooking appliances) there should be properly sited working CO detectors.

Indications of disrepair to any gas, oil, or solid fuel heating appliances, or defects to the flue, can often lead to staining around the junction between the appliance and the flue. Inadequate ventilation into the dwelling, or too powerful extract ventilation, can prevent sufficient air for combustion. The outlet to any flue should be sited away from fixed vents or openable windows.

Open-flued or flueless gas or oil heaters should not be present within the dwelling. Ideally, where there is a gas cooking hob or oven, there should be a hood and extractor venting to the outside. In addition there should be a ventilated lobby between an integral garage and the dwelling.

Where there are indications that there may be potential threats, following a visual inspection of the appliances and the means of ventilation, further investigation and a safety report from an appropriate engineer should be commissioned.

(NB: In 2019 the UK government made proposals to introduce a ban on gas-fired appliances for heating and cooking in new (yet to be built) dwellings, but no proposals for the removal/replacement of those in existing dwellings.)

Uncombusted fuel gas – As mentioned in Part 4, mains (or natural) gas is the most common gas used in urban and semi-urban areas. However, in rural and isolated areas it will be liquefied petroleum gas (LPG). Mains gas is primarily methane and is less dense than air, while LPG is denser than air.

As discussed for carbon monoxide, etc., above, p. 67), to avoid leakages appliances should be properly designed, installed, and regularly serviced and tested by a competent person. Any suspected defects should be dealt with.

If mains gas is being used, then (trickle) ventilation should be provided at a high level, but where LPG is used then there should be low-level ventilation.

If there is any indication that there may be a leak, the supply should be turned off at the main shut-off valve, and all naked flames extinguished.

Volatile organic compounds (VOCs) – There are two potential sources of VOCs: first, materials containing VOCs used in the construction of the dwelling; and, second, products introduced into the dwelling by the occupiers. In many countries, there are now controls on VOCs in household furniture and fittings, etc.

Generally, VOCs from materials used in construction will give off gas over a relatively short period, although this supports the idea that newly built dwellings should be well ventilated for (say) two or three weeks before occupation. Care should also be taken to ventilate well following any alterations.

Lead dust – Lead-based paint was widely used in dwellings until around 1970, and, where this remains, it can be a potential threat to health, particularly where it is flaking.

If lead-based paint is to be removed, then proper precautions should be followed to avoid ingestion of airborne lead particles, and to prevent the lead deposits within the dwelling.[17] This may necessitate specialist contractors and the temporary rehousing of residents.

Unlike other non-biological pollutants, it is possible to sample both dust and paint to check for the presence of lead. One method for testing for the presence of lead-based paint is using a portable (handheld) X-ray fluorescence (XRF) analyser, which avoids sampling the paint.

Radiation – This is primarily radon gas, and the UK (along with several other countries) has produced maps to show areas where high levels of radon may be found.[18] If a dwelling is in an area where radon levels are designated high, then it is possible to test using an accredited test kit.

If test results show high levels within the dwelling then various actions can be taken to reduce the levels or avoid radon penetration. The extent and nature of the work involved will depend on the construction of the dwelling.

Assessing space

There are two aspects to the assessment of space. One, that the size and the layout of the dwelling itself is adequate and appropriate for the size of household the dwelling has been designed to accommodate. And, two, that the dwelling is adequate and suitable for the household currently in occupation.

Assessment of the dwelling should cover whether there are sufficient rooms (including a kitchen or food preparation area, and sanitary and bathroom compartments), and whether those rooms and areas are of sufficient size to enable them to be used for the intended purpose.

Assessing whether the dwelling is adequate and suitable for the occupying household is relatively straightforward, requiring a comparison of the dwelling size and layout with the composition of the household (including the numbers, gender, and relationship of the household members).

The dwelling itself may be satisfactory, but not suitable for a household of a particular composition – i.e., the two are incompatible. The solution should focus on identifying alternative, suitable accommodation for the household.

Assessing noise

There should be a visual examination, taking account of the location of the dwelling (such as sited adjacent to roads or industrial premises), and if a flat or apartment where there could be noise from another dwelling or communal space is located nearby. As noise is likely to be more of a problem at night, a night-time visit may be necessary. Measurement of noise levels using properly calibrated noise meters can be helpful to confirm the subjective assessment. Such measurements may need to be carried out over a period of time. Noise includes vibration, and BS 8233: 2014 provides guidance and recommendations for the control of noise in and around buildings. It is necessary to ensure that measurements are carried out correctly.[19]

Assessing fire safety

As discussed in Part 4, there are several factors that could increase the risk of fire. First is ignition, which is often a result of occupier activities. But, the state (and age) of the electrical installation may need to be checked and tested by an electrical engineer. A possible problem could be a lack of sufficient power outlets (sockets), particularly considering the number of electrical appliances; overloaded sockets and trailing wires can increase the risk of a fire and can be trip hazard.

If a fire starts, then it should be contained within the room in which it started. This means the walls and ceiling should be free from cracks (that could allow smoke and fire to pass through), and the door should be fitted closely to the frame. A close-fitting fire door should separate the kitchen area from the rest of the dwelling.

There should also be a working smoke/fire alarm to warn the occupiers of the fire, and a means of escape. For high-rise residential buildings there is particular advice given in the Ministry of Housing, Communities and Local Government (MHCLG) HHSRS Fire Addendum.[20]

Assessing dangerous design features – There are a range of features that can increase the likelihood of a physical injury in dwellings, particularly increasing the risk of falls (see Part 4, p. 37). Stairs are a feature that are a risk (even when well designed and in good condition), but the risk of a misstep and fall is increased where the dimensions change (particularly the height of the risers). Handrails are primarily a safety measure (to be grabbed to avoid a fall after a misstep), and research has shown that one handrail is better than none, and two handrails are better than one.[21]

There should be sufficient space in bathrooms and sanitary accommodation to enable the users to manoeuvre easily. There should be grab handles for baths and showers.

Unexpected (or ill-lit) ill-defined changes in floor levels within the dwelling can increase the likelihood of a misstep and fall. This again will be exacerbated by inadequate or inappropriate lighting and should be part of the assessment.

To limit the possibility of falls from landings (and stairs) there should be guarding to at least 1,100 cm, designed so as not to encourage climbing and with any openings less than 100 cm. To reduce the chances of small children falling out of windows, there should be child safety catches, restrictors (to limit the opening to less than 100 cm, and/or guarding on the outside.

Notes

1 https://www.childrensmercy.org/in-the-community/healthy-home-program/.
2 https://environnement.brussels/etat-de-lenvironnement/archives/rapport-2007-2010/environnement-et-sante/pollution-interieure-le.
3 http://www.cmei-france.fr/revendeurs.php.
4 https://www.weather.gov/key/climate_heat_cool.

5 https://www.ashrae.org/technical-resources/bookstore/standard-55-thermal-environmental-conditions-for-human-occupancy.
6 https://www.iso.org/standard/39155.html.
7 https://www.cibse.org/Knowledge/knowledge-items/detail?id=a0q2000000 8I7f5AAC.
8 EU Directive 2010/31/EU.
9 Ezratty V et al. (2017) Fuel Poverty in France: Adapting an English Methodology to Assess Health Cost Implications. *IBE* 26(7): 999–1008.
10 Ormandy D (ed.) (2009) *Housing and Health in Europe: The WHO LARES Project*. Routledge, Oxon.
11 https://www.statistiques.developpement-durable.gouv.fr/enquete-performance-de-lhabitat-equipements-besoins-et-usages-de-lenergie-phebus.
12 Ormandy D, Ezratty V (2012) Health and Thermal Comfort: From WHO Guidance to Strategy. *Energy* Policy 49: 166–121.
13 https://www.gov.uk/guidance/standard-assessment-procedure.
14 https://www.bre.co.uk/filelibrary/Briefing%20papers/117106-Assessment-Protocol-v2.pdf.
15 https://www.designingbuildings.co.uk/wiki/Damp_in_buildings.
16 https://www.who.int/airpollution/guidelines/dampness-mould/en/.
17 https://www.gov.uk/government/publications/advice-on-lead-paint-in-older-homes;https://nchh.org/information-and-evidence/learn-about-healthy-housing/lead/.
18 https://www.ukradon.org/information/ukmaps; https://radonkit.ca/radon-gas-map-for-canada-potential-radon-levels-across-canada/.
19 Colthurst A and Fisher S (2017) Noise and Vibration. In Battersby SA (ed.), *Clay's Handbook of Environmental Health,* 21st edn. Routledge, Oxon, pp. 879–935.
20 MHCLG (2018) *Housing Health and Safety Rating System Operating Guidance*. Addendum for the profile for the hazard of fire and in relation to cladding systems on high rise residential buildings. https://assets.publishing.service.gov.uk/government/uploads/system/uploads/attachment_data/file/760150/Housing_Health_and_Safety_Rating_System_WEB.pdf.
21 Roys M (2013) *Refurbishing Stairs in Dwellings to Reduce the Risk of Falls and Injuries*. BRE Trust, Watford, Herts.

Part 7

Conclusions and perspectives

Housing is a social determinant of the health and well-being of residents. Living in a dwelling that fails to meet the needs of occupiers has potentially serious consequences for health, well-being, and life chances. Dwellings should provide a safe and healthy environment; providing shelter where the household can relax after work and school, be a place to enjoy family life and feel safe; and should allow (and encourage) connection with the community, promoting social inclusion. Dwellings should not have a prejudicial impact on physical and mental health, rather they should have a positive effect.

This monograph has set out the background and the research and evidence that informs the standards and the assessment of conditions, and has discussed examples of how standards can be formulated. However, standards are purposeless without a system, a procedure, to apply them and to ensure compliance.

Assessments

The first stage is to assess dwellings to determine whether they comply with the relevant standard(s). Assessments can be made by a generalist, or by separate specialists. In the UK, assessments are generally made by a qualified generalist – an environmental health practitioner (EHP). Where the EHP identifies a problem that is complex or where the cause is unclear, the EHP should call in a specialist to complete the assessment and to recommend solutions. This is often the case for potential electrical or gas faults, for problems of energy efficiency, or for suspected structural failure.

Relatively few countries have generalists, relying on specialists who deal only with a particular issue, such as lead dust and lead-based paint, carbon monoxide, or energy efficiency. It is arguable that an all-inclusive approach using a qualified generalist is more efficient, recognising that when it comes to providing a healthy home environment, a dwelling is more than the sum of its parts. It also avoids the possibility of solving one problem while ignoring or even exacerbating another. Finally, it is probably a more cost-effective way of ensuring that all deficiencies that could impact on the health and safety of occupiers are identified and remedied. It also means that those with concerns about housing conditions only have to contact one office or officer.

As has been made clear, environmental health practitioners make a unique contribution to public health through their problem-solving skills, which allow them to intervene in the causes of ill health in the dwelling.[1] They can only do this with the support of their employing authorities, the government, and other agencies, given that action to reduce the harmful impacts of housing conditions has the potential to save costs for the health services and society generally.

Ensuring compliance

Where it has been determined that a dwelling has failed to meet the requirements of a standard, the question then is how to ensure the necessary steps are taken to remedy the problem or deficiency. There are several answers, depending on whether the dwelling is rented or is owner-occupied.

While this monograph concentrates on existing dwellings, it is worth mentioning the control of the construction of new dwellings and dwellings subject to major alterations.

New dwellings – Building regulations (or building codes) give standards and requirements for the construction of all buildings, including dwellings and building works to existing structures. These focus primarily on the structure and the safe and proper fitting of facilities. The main purpose is to protect safety and welfare by ensuring there is compliance with the regulations (or codes).

Checking for compliance will involve checking those elements of the structure that will be covered over as building work progresses (such as foundations and drains).

There are two approaches – first, the inspection and certification by a specialist officer from the local authority (or state agency); and second, a form of "self-certification", either by the constructor, or by a person appointed by the constructor.

The first of these provides an independent assessment and certification. In England and Wales, where the local authority officer decides that the construction work does not comply with the building regulations, the local authority has powers to require the "defective" construction be put right and/or to prosecute. The second approach relies on trust. As the certification is detached from the local authority, whether or not an appointed person considers there is non-compliance, and an informal approach to the constructor does not solve the problem, then the matter should be reported to the local authority for further action.

As has been demonstrated with the fire at Lakanal House in the UK (July 2009) and other fires in apartment blocks around the world, apparent compliance does not always guarantee that work has been carried out properly and that the premises are safe. The pressure on de-regulation and "self-certification" can lead to inadequate control of the building work on site.

Rented dwellings – There are two approaches aimed at dealing with unsatisfactory conditions. First, public intervention and, second, private action for breaches of the tenancy contract.[2]

A local housing authority (or its equivalent) is responsible for ensuring rented dwellings (both privately owned and non-council-owned) satisfy the relevant standard. In the UK, an inspection and an assessment are carried out by an officer of the local housing authority, usually an environmental health practitioner, a qualified generalist. Where necessary, a specialist will be called in to advise, both on the assessment and the solution. If the assessment identifies conditions that are a potential threat to health and/or safety, the local authority either must or can (depending on the severity of the threat) take action to require remedial works. (What those remedial works are will depend on the construction of the dwelling and the nature of the problem.)

The enforcement procedure in England and Wales is (unnecessarily) complex and time-consuming and gives priority to the rights of the landlord over and above those of the tenant.[3] The landlord is given a right to appeal (question) the action taken, while the tenant remains under an obligation to pay rent to continue the right to remain in occupation of the dwelling (a dwelling that has been judged to be a threat to health). In France, the balance is slightly different. Where action is taken to require remediation of "insalubrious" conditions, the obligation to pay rent is suspended (and so is the tenancy time period) until the remedial action has been completed.

Tenancy contracts (agreements) will include obligations on the landlord to maintain the dwelling in a satisfactory condition (however phrased). Where the landlord fails to do so, the tenant has the option to take legal action for failure to meet the obligations (a breach of contract). This assumes two factors – that the tenant can afford to take legal action, and that the tenant has some security of tenure (is protected against eviction). Neither of these is certain in the UK.

Owner-occupied dwellings – Standards, assessments, and the enforcement legislation are usually tenure-neutral – i.e., they apply to all dwellings irrespective of ownership. Although housing is a national asset, individual dwellings are also personal property (a financial asset) of the owner, and there is a reluctance to take legal action against owner-occupiers. It is however possible, and may in some circumstances be appropriate (perhaps to protect the householder or members of the household).

Who pays?

Usually it is the owner of the dwelling who pays for the cost of any works. However, there is a range of grants (financial assistance) and loans available, each with their own eligibility criteria. The type of works that may qualify for a grant or loan also varies – they may cover works of repair, general improvements (modernising), energy efficiency improvements, or adaptations for a disabled occupier.

Most grants or loans are limited to owner-occupiers, although some may be available for landlords and, in some cases, tenants (with the consent of the landlord).

Giving public money (or loaning it at favourable interest rates) to the owner of a dwelling may initially seem illogical (after all, the dwelling is a financial asset belonging to the owner). However, there is evidence to show that removing threats to health and safety in dwellings results in cost savings to the health sector, benefits the local and national economies (reducing days off work), and benefits potential educational achievements (reducing days off school), so improving life chances.[4] (NB: The calculations relating to the cost savings to the health sector were made possible by the English Housing Health and Safety Rating System, as it links specific housing conditions with health outcomes.[5,6])

And finally

Research has provided a burgeoning bank of evidence on the relationship between housing conditions and health, and this research is continuing to do so. This evidence gives a solid basis for housing standards and guidelines. But, without the qualified generalists, without the specialists, and without a mechanism for the effective application and implementation of those standards and guidelines, plus the necessary resources (and political commitment), they are rendered useless.

Notes

1 Battersby S (2017) Historical Context, Philosophy and Principles of Environmental Health. In Battersby S (ed.), *Clay's Handbook of Environmental Health*, 21st edn. Routledge, Oxon. pp. 1–60.

2 Burridge R, Ormandy D (2007) Health and Safety at Home: Private and Public Responsibility for Unsatisfactory Housing Conditions. *Journal of Law and Society* 34(4): 544–566.

3 Ormandy D, Battersby S (2019) Landlords' Rights Trump Public Health. *Journal of Housing Law* 22(2): 28–35.

4 Marmot Review Team (2011) *The Health Impacts of Cold Homes and Fuel Poverty*. Friends of the Earth and the Marmot Review Team, London. Examined on 8 May 2018 at https://friendsoftheearth.uk/sites/default/files/downloads/cold_homes_health.pdf.

5 Roys M, Davidson M, Nicol S, Ormandy D, Ambrose P (2011) *The Real Cost of Poor Housing*. BRE FB 23. IHS BRE Press, Bracknell; and subsequent BRE Reports.

6 Ezratty V et al. (2019) Health Cost Benefits of Energy Upgrades in France. In Jones M, Rice R, and Meraz F (eds). *Designing for Health & Wellbeing*. Vernon Press, Malaga.

Index

For Product Safety Concerns and Information please contact our EU
representative GPSR@taylorandfrancis.com Taylor & Francis Verlag GmbH,
Kaufingerstraße 24, 80331 München, Germany

Batch number: 08153772

Printed by Printforce, the Netherlands